Imperial Hygiene

A Critical History of Colonialism, Nationalism and Public Health

Alison Bashford
*Vere Harmsworth Professor of Imperial and Naval History,
University of Cambridge, UK*

© Alison Bashford 2003, 2014

All rights reserved. No reproduction, copy or transmission of this publication may be made without written permission.

No portion of this publication may be reproduced, copied or transmitted save with written permission or in accordance with the provisions of the Copyright, Designs and Patents Act 1988, or under the terms of any licence permitting limited copying issued by the Copyright Licensing Agency, Saffron House, 6–10 Kirby Street, London EC1N 8TS.

Any person who does any unauthorized act in relation to this publication may be liable to criminal prosecution and civil claims for damages.

The author has asserted her right to be identified as the author of this work in accordance with the Copyright, Designs and Patents Act 1988.

First published 2003
First published in paperback 2014 by
PALGRAVE MACMILLAN

Palgrave Macmillan in the UK is an imprint of Macmillan Publishers Limited, registered in England, company number 785998, of Houndmills, Basingstoke, Hampshire RG21 6XS.

Palgrave Macmillan in the US is a division of St Martin's Press LLC, 175 Fifth Avenue, New York, NY 10010.

Palgrave Macmillan is the global academic imprint of the above companies and has companies and representatives throughout the world.

Palgrave® and Macmillan® are registered trademarks in the United States, the United Kingdom, Europe and other countries

ISBN: 978-1-403-90488-1 hardback
ISBN: 978-1-137-42921-6 paperback

This book is printed on paper suitable for recycling and made from fully managed and sustained forest sources. Logging, pulping and manufacturing processes are expected to conform to the environmental regulations of the country of origin.

A catalogue record for this book is available from the British Library.

A catalog record for this book is available from the Library of Congress.

Transferred to Digital Printing in 2014

For Carolyn

Contents

List of Figures	ix
Acknowledgements	xi
Preface to the Paperback	xiv
List of Abbreviations	xvi
Introduction: Lines of hygiene, boundaries of rule	1

1. Vaccination: Foreign bodies, contagion and colonialism — 14
   - Foreign bodies: boundaries and the logic of vaccination — 16
   - Connections: empire and the genealogies of vaccine — 23
   - Vaccination and administration: certificates, scars and passports — 33

2. Smallpox: The spaces and subjects of public health — 39
   - Smallpox in Sydney, 1881 — 40
   - Lines of hygiene: bodies in quarantined space — 45
   - The carceral spaces of public health: government, consent and the liberal subject — 51

3. Tuberculosis: Governing healthy citizens — 59
   - Tuberculosis prevention and treatment: toward 'public' health — 63
   - Isolation and the dangerous consumptive — 66
   - The sanatorium: the cultivation of healthy selves — 70
   - Hygienic citizenship — 77

4. Leprosy: Segregation and imperial hygiene — 81
   - 'An Imperial Danger': contagion and segregation — 83
   - Exile-enclosure: island isolation in Australia — 93
   - Racial *cordons sanitaires* — 103
   - Interior frontiers: sexuality, contact and race — 107

5. Quarantine: Imagining the geo-body of a nation — 115
   - Quarantine and nationalism — 117
   - The island-nation: marine hygiene and the national border — 123
   - Imagining Australia in space and time — 129

6. Foreign bodies: Immigration, international hygiene 137
   and white Australia
   - International hygiene 141
   - Racial imaginings and white Australia 147
   - Imperial migration and racial hygiene 151
   - Tropical medicine and foreign white bodies: 157
     'Is White Australia Possible?'

7. Sex: Public health, social hygiene and eugenics 164
   - Venereal disease: detention and education 166
   - Reproduction and responsibility 172
   - The eugenic *cordon sanitaire*: contagion and the 180
     future population

Conclusion 186

Notes 190

Select bibliography 232

Index 246

List of Figures

Figure 1.1	Title page to *Animal Vaccination*, Thomas Richards, Sydney, 1882, p. 2. Courtesy Mitchell Library, State Library of New South Wales.	26
Figure 1.2	Animal Vaccination as practised at Bombay. From *Animal Vaccination*, Government Printer, Sydney, 1882. Courtesy, Mitchell Library, State Library of New South Wales.	27
Figure 1.3	Return of Patients Inoculated, 1802. Thomas Christie, *An Account of the Ravages Committed in Ceylon by Small-Pox*, J.&S. Griffith, Cheltenham, 1811. Courtesy, The Wellcome Library for the History and Understanding of Medicine, London.	31
Figure 1.4	Smallpox. From J. Ashburton Thompson, *A Report to the President of the Board of Health containing photographs of a person suffering from variola discreta, and accounts of the case*, Government Printer, Sydney, 1886. Copy in author's possession.	37
Figure 2.1	Vaccination. *Illustrated Sydney News*, 9 July 1881, p. 12. Courtesy, The State Reference Library, State Library of New South Wales.	47
Figure 4.1	Nationality of recorded cases of leprosy in Australia. Adapted from J.H.L. Cumpston, *Health and Disease in Australia* (1928) introduced and edited by Milton Lewis, Australian Government Publishing Service, Canberra, 1989, p. 209.	95
Figure 4.2	Leper Pedigree. From J. Ashburton Thompson, 'A Contribution to the History of Leprosy in Australia', in *Prize Essays on Leprosy*, New Sydenham Society, London, 1897, p. 159. Courtesy, the State Reference Library, State Library of New South Wales.	97
Figure 5.1	World Map. W. Perrin Norris, *Report on Quarantine in Other Countries and on the Quarantine Requirements of Australia*, Government Printer, Melbourne, 1912. Courtesy, Mitchell Library, State Library of New South Wales.	127
Figure 5.2	Isochronic Chart, from *Health*, 5 (1927): 46. Courtesy, Mitchell Library, State Library of New South Wales.	132

x *List of Figures*

Figure 7.1   Pre-marital screening.   175, 176
and 7.2   Racial Hygiene Association, *Annual Report*, 1938–39.
                 Courtesy, Mitchell Library, State Library of New
                 South Wales.
Figure 7.3   Sow the Seeds of Good Health.   179
                 New South Wales Department of Health Poster c.1930.
                 Courtesy, The Powerhouse Museum, Sydney.

Acknowledgements

Neither by temperament nor by training am I a local historian. Entirely fortuitously, however, much of this book has been written over the past year on isolated North Head, Sydney Harbour, in my house on remnant army land soon to be National Park. The old Quarantine Station is my immediate neighbour on one side, the equally out-of-use Catholic seminary on the other side. The military fort lies on a third and to the east are huge cliffs dropping to the ocean. For years the telling historical geography of this liminal bit of metropolitan Sydney had been the object of my research – see details in Chapter 2. And for the past year, I have been able to take breaks from writing, and explore along the stone walls that criss-cross the headland in unexpected ways – walls which used to separate the first class from the steerage class immigrants-in-quarantine, the whites from the Chinese, the plague-victims from the army, the infected and the carriers from the dead, and everyone from the priests-in-training. The sanitary lines between *Purity and Pollution* began to interest me in my first book, and here I have both lived in and thought much about related, but different kinds of nineteenth and twentieth century delineations between the 'hygienic' and the 'unhygienic', the 'fit' and the 'unfit': imperial lines of racial segregation, lines of defence, lines of nation.

This writing time-and-place has been shared with important people in my extended family, and my thanks to them all: my twins Tessa and Oscar, Carolyn, Catriona, Nicholas. The Bruntons – especially Nicholas and Carol – have been generous and full of grace. My parents continue to provide every kind of support and interest.

I have been writing versions of these chapters for several years now. My thanks to readers of various drafts: Barbara Caine, Harriet Deacon, Catriona Elder, Stephen Garton, Mark Finnane, Bernard Harris, Claire Hooker, Philippa Levine, Roy Macleod, Renisa Mawani, Dirk Moses, Glenda Sluga, Carolyn Strange and Mick Worboys. I am grateful especially to Renisa and Mick for the opportunity of reading their unpublished essays on leprosy, from which I have learned a great deal. Any residual mistakes or (what might be considered) strange interpretations in the book of course remain my own. Other friends and colleagues have offered suggestions and advice, or have listened to ideas and contributed generously from their own expertise. My thanks to Warwick

Anderson, Adrian Carton, Yvonne Cossart, Joanne Finkelstein, Caroline Jordan, Fiona Nicoll, Maggie Pelling, Milton Lewis, Penny Russell, and Anna Yeatman. My brother Guy Bashford has been both interested and – it has to be said – useful in offering specialist medical details and information – my thanks. Archivists and librarians at the Mitchell Library in Sydney and the Wellcome Library for the History and Understanding of Medicine, London, have, as always, been endlessly accommodating. In particular, thank you to Lesley Hall.

Claire Hooker, Martha Sear, and Megan Hicks of the Powerhouse Museum, Sydney, have discussed public health with me for many years now, and I have learned much from them. We undertook a collaborative research project, some of which informs these chapters. Since it is not quite appropriate for co-editors to praise and thank each other in the acknowledgements of edited collections, I can take the opportunity here to thank and praise Claire Hooker – my co-editor of *Contagion* and sometime co-author. Her impeccable research assistance runs through many of these pages. But I have benefited even more from her intellectually sharp and challenging questions and commentary. Other research assistants have intermittently been brought on board, and I am very grateful indeed for their help and interest: Michelle Arrow, Sarah Howard, Maria Nugent, Marijana Vurmeska, and for many years, Megan Jones.

Carolyn Strange has had more discussions about this book than anyone could, or should ever reasonably expect. Luckily for me, she is not only generous with her time and remarkable capacity to discuss others' historical ideas, but she is also the best editor I know: all residual passive constructions remain my own. I hope the result is some testimony to all the conversations and all the reading over these years. My deepest thanks.

I have also gained a great deal from exchanging ideas with colleagues at seminars and conferences. In particular, I am grateful to friends in the History Department, the Gender Studies Department and the Unit for the History and Philosophy of Science at the University of Sydney who have heard many versions of these chapters; the National Centre for HIV-Social Research at the University of New South Wales, where early ideas were clarified for me; the 1999 Berks at Rochester which was most stimulating; the Social History of Medicine conference on Tuberculosis in 2001 organised by Flurin Condrau and Mick Worboys; and most recently, the workshop which Philippa Levine organised in LA for contributors to the *Oxford History of the British Empire* – was the Empire ever so interesting as at that meeting? I have received two generous Australian Research Council grants which have enabled precious research and writing time, and for which I am extremely grateful. Travel grants from the Wellcome Trust and the University of Sydney have facilitated research in London and attendance at these conferences. In 1999 I spent time at the Institute for Advanced Studies in the Humanities, University of Edinburgh, which offered every possible academic support. My thanks to Fiona Probyn for taking on my teaching duties with such skill during these periods.

The arguments of some of the chapters have been sketched in previously published work. All of the chapters here are major revisions of these articles, reframed in terms of the argument of the book and supplemented in terms of primary material and recent secondary literature: 'Quarantine and the Imagining of the Australian Nation', *Health* 2 (1998): 387–402; 'Epidemic and Governmentality', *Critical Public Health* 9 (1999): 301–16; 'Foreign Bodies', in Alison Bashford and Claire Hooker (eds), *Contagion* (Routledge, 2001); 'Leprosy and the Management of Race, Sexuality and Nation', in Alison Bashford and Claire Hooker (eds), *Contagion* (Routledge, 2001), co-authored with Maria Nugent and drawn on here with her permission; and 'Cultures of Confinement', in Carolyn Strange and Alison Bashford (eds), *Isolation* (Routledge, 2003). I am grateful to the Powerhouse Museum for permission to reproduce the poster, and to the State Library of New South Wales for permission to reproduce illustrations and maps.

Alison Bashford,
December 2002

Preface to the Paperback

The paperback of *Imperial Hygiene* appears one decade after its first publication. It is an anniversary edition of sorts. At the very least, this edition affords me the opportunity to look back on its claims in the light of ten years' reading and conversation, just the right amount of space and time for reflection on an intellectual endeavour. This was my second book, and I noted in the Acknowledgements how its research questions derived from the spatial sensibility that had shaped its antecedent, *Purity and Pollution*. In turn *Imperial Hygiene* went on to inspire a number of successor studies, mainly collaborations: a book on global eugenics with Philippa Levine and on geographer Griffith Taylor with Carolyn Strange. And even as *Imperial Hygiene* was appearing on the shelves in 2003, *Medicine at the Border* was in the making, already comprehended as its twin. The second-born, it emerged after a good interval, as all the best twins do.

The Foucauldian drive of *Imperial Hygiene* marks a particular moment in history-writing. Both historians of medicine and spatially-inclined historians were served well by the great wave of scholarship over the 1990s. Clearly the sex-geography-government triad that captured Michel Foucault's imagination was important for rethinking disease and its spatial management. It opened up particular possibilities through which the improvement imperative of public health could be thought through as the production of racialised, gendered and civic selves. How this happened in spaces of different scales was what caught my imagination at the time. The fortuitous element was my geographic location in Australia, a national context that distilled questions of health, citizenship, race and disease control with particular intensity. Even more fortuitous has been the thriving public commitment to humanities research in that country, in the form of the Australian Research Council. I remain indebted to this remarkable institution, and to the rigorous system of intellectual assessment through which it functions.

In the intervening decade, my thoughts on population and health in colonial Australia set down in *Imperial Hygiene* turned outwards, a turn common to many historians of my generation. We were hardly the first to think globally. Nonetheless, *Imperial Hygiene* caught the beginning of an expansive moment in transnational historical writing. At the same time, this book stemmed from a history of medicine/history of science tradition that had long traced ideas and practices across oceans and continents. This made the "global turn" look slightly less novel than many practitioners then claimed. In any case, I have since made the turn in a committed way, seeking to connect history of medicine with world history, and, continuing the geographic links, with environmental history. The spatially-inflected biopolitics of *Imperial Hygiene* has informed the bio-inflected geopolitics of my most recent book, *Global Population*.

I acknowledge the Powerhouse Museum, Sydney, for permission to reproduce the cover image, The Wellcome Trust for permission to reproduce Figure 1.3, and the State Library of New South Wales for permission to reproduce the remaining Figures. My thanks to Jenny McCall, Clare Mence and Emily Russell at Palgrave for inviting and producing a paperback of *Imperial Hygiene*. Except for corrections of errors, the book is published here unchanged. Quarantine history continues apace in Sydney, made all the more interesting and enjoyable for the company of colleagues Annie Clarke, Ursula Fredrick and Peter Hobbins, as we pursue our project on the inscriptions on North Head, Sydney Harbour. That place, home for so many years, remains the local site that inspires a global history.

AB
Sydney and Cambridge, November 2013

List of Abbreviations

BMJ	The British Medical Journal
Cth	Commonwealth (of Australia)
ML	Mitchell Library
NAA	National Archives of Australia
NH&MRC	National Health and Medical Research Council
NSW	New South Wales
Qld	Queensland
QSA	Queensland State Archives
QHSO	Queensland Home Secretary's Office
Vic.	Victoria

Introduction: Lines of hygiene, boundaries of rule

'Imperial cleanliness,' wrote an early twentieth century public health bureaucrat is 'development by sanitation ... colonising by means of the known laws of cleanliness rather than by military force'.[1] Like many of his contemporaries, the connection between hygiene and rule was obvious for this commentator, both commonplace and a driving mission. This relationship between public health and governance has, in many ways, been rediscovered by critical sociologists and historians of health and medicine. 'The power to govern', wrote one 'is often presented as the power to heal'.[2] This book is about the historical relatedness of public health and governance, hygiene and rule over the nineteenth and twentieth centuries, taking as case studies the British colonies in Australasia, and subsequently Australia itself as a white 'settler' society, as a colonising nation. The book is about the enclosures, boundaries and borders which were the objects and means of public health, as well as of colonial, national and racial administration. Lines of hygiene *were* boundaries of rule in many colonial and national contexts.[3] Conversely – and this is perhaps the more novel contribution – many boundaries of colonial rule manifested as and through lines of hygiene, as spaces of public health. Imperial Hygiene traces all kinds of public health spaces, and explores their intersection and oftentimes neat dovetailing with other governmental 'lines', other real boundaries of rule: national borders, immigration restriction lines, quarantine lines, racial *cordons sanitaires* and the segregative ambitions of a grafted eugenics and public health. All these spaces – these therapeutic, carceral, preventive, racial and eugenic geographies – produced identities of inclusion and exclusion, of belonging and citizenship, and of alien-ness.

As it developed over the nineteenth century, public health was in part a spatial form of governance. The knowledges, institutions and

1

practices of public health aimed to regulate the circulation of matter (or people) constituted as dangerous *because of* their circulation and contact with unknown people in unknown places: prostitutes with venereal diseases, waterborne microbes in the drains and sewers connecting urban spaces, infected migrants, soldiers or seamen travelling the globe, smallpox accompanying trading routes. Lines or barriers drawn *across* these global, local and bodily circulations and connections are often what have constituted public health measures: *cordons sanitaires* of various kinds. But if public health was in part about segregation (of the diseased from the clean, the fit from the unfit, the immune from the vulnerable), so was race a segregative discourse. This book is a contribution to a growing historical scholarship that seeks to examine segregation as both hygienic – that is, as part of public health – and racial – as part of the systems and cultures of race management, including as I emphasise throughout, the management of whiteness.[4] This is a specific trajectory of the now extensive field of scholarship on medicine and colonialism on the one hand,[5] and of the less developed field of medicine and nationalism (including National Socialism) on the other hand.[6]

In broad terms, the book covers the period between the mid 1860s and the Second World War. The period can been delineated in several ways: in terms of medicine and public health; in terms of government; and in terms of racial ideas and practices. In medical and public health terms, this was the period in Britain and Australia of an expanding scope and bureaucratisation of public health within government. It also bounds the Imperial Vaccination Acts and the Contagious Diseases Acts around the 1860s and the general uptake of antibiotics and mass immunisation after the Second World War, the efficacy of which shifted the main problem of public health away from communicable diseases, and towards the prevention and management of chronic conditions. The period has also been described as the racial century,[7] the consolidation and rigidifying of racial categories in the second half of the nineteenth century, and what I call the eugenic half-century which followed. The postwar period was marked by decolonisation (alongside the instruments and institutions of 'world health') and a critical reassessment, at many levels, of western nationalism and racism: the explicit connection of nation, race and hygiene common to Britain, its dominions, and to Nazi Germany became almost unspeakable after the Second World War. Spanning the 1860s to about 1950, the book also traces some of the changes in the relation between public health and shifting modes of liberal rule: from classical Victorian liberalism and its transplantation to the colonies, and the problems like compulsory vaccination and public health detention for liberal governance, to the developing social government and welfare of the early to mid twentieth century. Again, after the War, there was a shift in liberal governance towards 'advanced liberalism'.[8] While the study of public health and advanced liberalism or neoliberalism of the later twentieth century has drawn considerable scholarly attention,[9] I turn back to the earlier generations with the aim of historicising this existing literature. I deal, then, with the late nineteenth century generation who inherited sanitary reform; the early twentieth century generation who saw 'welfare' (and the idea of 'health and welfare') emerge; and the interwar generation whose rigid, scientised and expert formations of 'nation', 'race' and 'health' intersected ever-more tightly with welfare itself and with conceptions of citizenship and exclusion from citizenship.

While referring to several colonial contexts (India, the Straits Settlements, Canada, Fiji) as well as to Britain itself, I focus most closely on the self-governing British colonies in Australasia, and, after federation of these colonies in 1901, the (colonising) nation of Australia. In its formative stage from the 1880s, Australian nationalism was based squarely on the idea of white Australia: it was explicitly a nationalism of race. The pursuit (at many levels) of health, hygiene and cleanliness was one significant way in which the 'whiteness' of white Australia was imagined, as well as technically, legally and scientifically implemented: purity was the project of public health, as well as the project of nation. In this way, questions of race and a racialised geopolitics structured and shaped the knowledge, practice and bureaucracy of public health in Australia in the same fundamental way that questions of class and urbanisation shaped British public health in the nineteenth century. The Australian colonies and nation emerge as exemplary sites through which to draw together the literature on race and public health segregation, with critical histories of white settler colonies and their complicated boundaries of rule, which have challenged and enriched colonial and imperial histories more generally.[10]

Australian history unsettles the categories in imperial and commonwealth historiography of 'colonisation' and 'settlement', or 'the protectorates' and 'the (white) dominions'. Just what one means by 'colonial' in Australian history always needs clarifying. In Indian history, for example, 'colonial' and 'postcolonial' is periodised at decolonisation, in however complicated a way. Australian history always asks one to define 'colonial': colonial as in British rule before the self-governing 'nation' of 1901, or 'colonial' as in the ongoing colonisation of

Indigenous people, for whom the distinction between 'British' white rule and the white rule of the new Commonwealth of Australia after 1901, is less than significant. From this perspective there is no decolonising moment which makes Australia post-colonial.[11] So, nationalism in Australian history is not anti-colonial nationalism in the sense usually taken up in postcolonial studies, but rather the nationalism of white Australians partially separating themselves from British rule, but at the same time thoroughly identifying as Britons, indeed defining their new nation as white *vis-à-vis* Asian, especially Chinese others. Taking Australia as a case-study then, exposes 'imperial' in several senses: the management of Indigenous populations, the modes by which British populations 'settled', and the nationalism produced through that 'whiteness' which was also Britishness.[12] This all involved complicated policies and practices of race and health management of Indigenous communities, of white Britons and their entry into the territory and the body politic, and of Asian people, who were deported and subsequently excluded in the pursuit of an imagined white (read: pure, clean, uncontaminated) Australia.

The pursuit of 'health' has been central to modern identity formation. It has become a way of imagining and embodying integrity and, problematically, homogeneity or purity of the self, the community, and especially in the early to mid twentieth century, the nation. Nation-forming has found one of its primary languages in biomedical discourse,[13] partly because of its investment in the abstract idea of boundary, identity and difference, but also because of the political philosophy that thinks of the population as one body, the social body or the body of the polity.[14] One result has been a cross-over of biomedical and politico-military languages of defence, immunity, resistance and invasion, of the body, the community and the nation.[15] My interest here is to explore not only the significance of the metaphor of the 'social body', but also the actual corporeal connectedness of bodies, communities and nations which were the business of public health. I look at substantive issues like arm-to-arm cowpox vaccination which literally connected populations of children across the Empire, the spread of communicable diseases, and the biological connections of sex, especially in the eugenic era when previously social attributes were refigured as inheritable, as 'contagious' between generations.

The idea of 'hygiene' connects all the chapters, and the substantive sites of analysis. From the late nineteenth century, and escalating until the Second World War, 'hygiene' came to be a personal and political imperative and mission, a noun which spawned ever-more adjectives which connected the bodily and the personal to larger governmental projects: sex hygiene, domestic hygiene, social hygiene, national hygiene, moral hygiene, tropical hygiene, maternal hygiene, racial hygiene, international hygiene and more. Hygiene was something which people could and did do to themselves and to each other: it was a practice. But it also had a significance greater than oneself. Victorian culture made 'cleanliness' into a subjectivity, a practice which shaped one's soul. By the interwar years, hygiene was a responsibility, a duty.[16] Hygienic imperatives were everywhere producing institutions and practices of public health which to be sure demonstrably reduced morbidity and mortality, and created what was for many a desirable order, cleanliness and safety. But what else produced, and was produced through this personal/national/imperial hygiene? In many ways, this question drives the book.

Signalling the constant need for purification from the ever-present contaminating threat over the border, however imagined, hygiene became a primary means of signification by which those borders were maintained, threats were specified, and internal weaknesses managed.[17] Far from being a straightforward metaphor, the use of the term 'hygiene', particularly in the context of nationalism, was a result of the deep connection between the political and cultural imagining of bodies and nations, as well as a long history of an 'imaginary geographics' of exclusion by which, as geographer David Sibley puts it, 'others' of all kinds were located 'elsewhere'. He writes 'This "elsewhere" might be nowhere, as when genocide or the moral transformation of a minority like prostitutes are advocated, or it might be some spatial periphery, like the edge of the world or the edge of the city'.[18] All kinds of edges, borders and peripheries are analysed here, as are all kinds of matter and people which crossed them: from the foreign body of cowpox matter entering the pure skin of the child which I examine in Chapter 1, to the foreign body of the unfit immigrant prohibited from the 'pure' nation, which is the subject of analysis in Chapter 6.

Anthropologists of colonialism have also been interested in identification and differentiation, those 'boundaries of rule' and 'colonial categories', which Ann Laura Stoler has richly analysed. These were often informal boundaries, what Stoler calls 'interior frontiers' (after philosopher Johann Gottlieb Fichte and theorist of nationalism Etienne Balibar) differentiating kinds of colonisers, settlers, Indigenous people, métis, half-castes. She writes: 'a frontier locates a site both of enclosure and contact and of observed passage and exchange. When coupled with the word *interior*, frontier carries the sense of internal distinctions within a territory (or empire); at the level of the individual, frontier marks the moral predicates by which a subject retains his or her national identity despite location outside the national frontier and despite heterogeneity within the nation-state'.[19] Such interior frontiers certainly interest me. I examine, for instance, the interior frontiers produced by white sexual and social conduct in the precarious zone of the tropics, and connections to the management of leprosy, often considered to be transmitted to the white population by sexual contact between races. By the same token, if 'making boundaries' (of race, citizenship, culture, and of legitimate and illegitimate sexual contact) has been the analytical draw of anthropologically inflected histories of colonialism,[20] I am drawn in the first instance to far more material lines, edges, borders, displacements and enclosures, and their constitution through and by public health. I examine for example threshold places like the policed fences in quarantine zones in smallpox epidemics, the momentary border between the diseased and the clean. In Chapters 5 and 6 I look at the new national borders of an island-nation which became a line of both quarantine inspection of people and goods, and a racialised immigration line. Internally, lines of race and hygiene criss-crossed the country. The 'Leper Line' for example was the 20th parallel, south of which Aboriginal people could not travel, under Western Australian law of the 1940s. Another version of segregation that I look at in Chapter 7 were early twentieth century lock hospitals for venereal diseases which doubled as native reserves. These kinds of geographic, legal and actual boundaries formed people's senses of identity and difference – they were about both administration and subjectification. Yet they were not separate from lines of the imagination: the boundaries, enclosures and segregations of eugenic, nationalist, imperialist and racist dreamers in the modern period; the great eugenic ambition of a clear line between the fit and the unfit; the aspiration of various Australian government medical officers of unambiguous classification and spatial and sexual separation of whites from Indigenous people, and 'half-caste' from 'full-blood' people; the fantasy line imagined by one geographer to be drawn across the island-nation, north of which should be Japanese, south of which should be British; or the lines of global racial distribution and differentiation drawn on so many maps of the world.

My (literally) grounding interest then, is in geographic, geopolitical, institutional and legal lines and segregations. My subsequent interest is in their intersection with interior frontiers, and with fantastic and utopian lines. Similarly, while I draw constantly from historical and anthropological studies on social contaminations and moral contagions,[21] I explore the problem of origin: communicable diseases and their management. This has led me to see afresh the extent to which public health and hygiene offered not just metaphors and rhetoric (or the 'pretexts' as Proctor writes of the Nazi enclosures of Jews in Warsaw)[22] for cleansing and purifying, but were the actual modes and tools of management for colonialism, nationalism, and in the interwar period, racial hygiene and eugenics: these were all part of the project and the imperative of public health.

Population health and liberal governance

There are two broad genealogies of public health: one is philanthropic-missionary, the other is governmental-political.[23] Broadly, they integrated during the regulating and bureaucratising years of the early to mid twentieth century, as government increasingly took responsibility for what had been charitable, philanthropic or parish concerns in the nineteenth century. In that this book is a history of technical and legal quarantines and communicable disease control as much as anything else, it is the government-political genealogy in which I am most interested: public health, state medicine, what was sometimes called social medicine. Much of the book takes up questions about how populations came to governed and managed through the rubric and problems of public health. I analyse not only the substantive issues or concerns (for example compulsory vaccination, maritime quarantine, health education, immigration restriction, isolation and segregation, eugenic pre-marital screening) but also the kinds of power exercised through these problems: coercion, consent, education, subjectification and power exercised through new conceptions of 'freedom'.[24] Public health was a field where techniques of liberal governance and authority were tried, resisted, abandoned, modified, outlawed, and normalised, techniques often based in, and departing from, penal systems and cultures. If Nikolas Rose has written of 'the liberal vocation of medicine', much work remains to be done on the precise history of this idea.[25]

Public health, 'population' health, is historically contemporaneous with, and part of, modern rationalities of government: political economy, liberal rule, nationalism, new politics of citizenship. In his synthesis of the history of public health, George Rosen emphasised the particular 'policy of power' and conception of society that underpinned the European mercantilist states. As Rosen summarised it, when policy makers asked the mercantilist question, 'what course must the government pursue to increase the national power and wealth?', they answered with reference to the size of the population, the material welfare of the population, and government control of the population.[26] Health, longevity, reproductivity, economy and security were in various ways and through various technologies, central. The pursuit of a greater health of the social body – the population – became a major aspiration of government from the mercantilist period, through the centuries of industrialisation, colonisation and liberalism. And 'population' as an idea, depended upon or required some sort of centralised state to be made meaningful.[27] Eighteenth-century developments in Utilitarian political philosophy also shaped the emergence of health and population as an imperative of government. Here Bentham was critical, his ideas underpinning so much social policy as it developed in Britain and the colonies through the nineteenth century, one line of which, through the great sanitarian and student of Bentham, Edwin Chadwick, became public health as we know it.[28] Both Bentham's perception that the welfare of the poor was an obligation of government, and the possibility that 'health' was necessary to 'happiness' and its pursuit, provided theoretical and moral principles on which all kinds of interventions were made, and expectations created, about the relation between government, the population and the individual.[29] As Dorothy Porter has succinctly summarised it: 'In the nineteenth century, public health reform interwove Victorian social science with Enlightenment political economy and was integrated into philosophical radicalism and the politics of social amelioration'.[30]

Foucauldian scholarship has formulated all this slightly differently to Rosen, as 'biopolitics'. Graham Burchell summarised biopolitics as a power which is 'exercised over persons specifically in so far as they are thought of as living beings: a politics concerned with subjects as members of a population, in which issues of individual sexual and reproductive conduct [I might add conduct in relation to health] interconnect with issues of national policy and power.'[31] The life of populations came to be seen as available for administration, indeed imperative to administer and manage in increasingly complex ways. By 'life' Foucault meant a discernible shift from the exercise of power through repressive punishment and threat of force and death, to the exercise of power through optimising the capacities of a population through interest in health, fecundity, illness, and longevity.[32]

Knowledge-techniques were developed or appropriated from other fields – in particular statistics – and were put to use as ways of gathering and formulating information about individuals and aggregating it into information about spatially defined, class defined, institutionally defined or sex defined abstract groups.[33] In the process, the very concept of 'population' – its political capacities, uses, limitations, and demands – was drawn into being. In the institutionalisation of the discipline of epidemiology over the nineteenth century, the central technology was statistics: 'the science of the state'.[34] And as Ian Hacking has written, 'the collection of statistics has created, at the least, a great bureaucratic machinery. It may think of itself as providing only information, but it is itself part of the technology of power in a modern state'.[35] The specific knowledges within the new human and social sciences and their tools – sociology, epidemiology, demography – flourished out of political economy, 'the science of wealth', and became increasingly refined as to their objects of inquiry over the nineteenth century. For Adam Smith, as Mary Poovey writes 'the working poor ... had to be conceptualized as an aggregate; because they could not govern themselves, they had to be governed from above'. A new 'social' domain was being delineated, the 'making of a social body'.[36] This view of both the economic significance of the working poor, but their simultaneous incapacity to govern themselves as the bourgeois subject could, explains the seeming paradox of liberal governance in the modern period, both distant and intervening, inclusive in theory, exclusive in practice. In colonial contexts this paradox was intensified.[37]

Colonial situations presented the immediate governmental problem of both the urban poor and 'ungovernable' Indigenous populations. Public health programmes and visions were a key way in which colonised people and territory were administered and came to be rendered intelligible to colonisers. Although differing vastly between colonies, and in terms of the responsibility governments took up for indigenous health, there was an enormous industry in the late Victorian and early twentieth century period in medico-administrative knowledge of indigenous people, as well as of white settlers.[38] Over time, indigenous people in many colonial contexts were brought into tighter relations of governance, of health and welfare and of sanitation.[39]

By historicising this medico-administrative knowledge and its effects, and interrogating particular problematisations determined by the geopolitics and race politics of Australian 'settler' colonialism and nationalism, I am writing a history of the colonial biopolitics of health. Colonial history needs to be integrated into the master narrative of what public health is and where it came from. This history was not supplementary to, but contemporaneous with and partly formative of, the consolidation of British public health over the nineteenth and twentieth centuries.

Lines of hygiene: from quarantine to the new public health?

Questions about how people have been governed, and rendered governable through and in the interests of public health have defined a field of critical medical sociology, and in particular the work of historically interested sociologists concerned with what has come to be called 'the new public health' of the later twentieth century. In one way or another the field derives from Michel Foucault's interest in dividing practices and spatial subjectification. Early systems of quarantine represented a fairly crude exclusion, what Foucault analogised as the treatment of the leper: 'a practice of rejection, of exile-enclosure; he was left to his doom in a mass among which it was useless to differentiate'.[40] This contrasted with the 'plague town' detailed in *Discipline and Punish* as a model of a new and qualitatively distinct kind of power.

> This enclosed, segmented space, observed at every point, in which the individuals are inserted in a fixed place, in which the slightest movements are supervised, in which all events are recorded, in which an uninterrupted work of writing links the centre and periphery ... all this constitutes a compact model of the disciplinary mechanism.[41]

Foucault argued that the social condition of epidemic produced early examples of disciplinary government, rendering subjects normalised through mechanisms of bodily training and self-surveillance.[42] I revisit versions of leper 'exile-enclosure' and 'plague towns' throughout the book. Although Foucault did not quite argue for the straightforward replacement of 'exile-enclosure' by the 'plague town', the implication of succession is there and the derivative field of scholarship carries this implication forward. Medical sociologist David Armstrong, whose work has been richly suggestive for my thinking through of public health, has devised a schema for the succession of regimes of public health over the last two centuries, which characterise differing – and evolving – modes of power.[43] Armstrong details the models of public health subsequent to the 'quarantine' model: sanitary science, personal hygiene and the 'new public health'. These became increasingly disciplinary in the Foucaudlian sense, until, in the new public health of the later twentieth century, danger is 'everywhere' and self-monitoring is more or less perpetual, but also re-worked as constitutive of selves. The (supposed) abandonment by western governments of crude isolation and quarantine responses in favour of more subtle and detailed public health strategies is often cited in critical sociological literature on health and medicine as a prime example of the historical shift away from the use of sovereign force, to the internalisation of a desire for health on the part of each citizen, 'the new public health'. Alan Sears argues for example that the *cordon sanitaire* shifted from the public to the diffused domestic realm in the early twentieth century. He writes that 'new-fashioned quarantine is not a blanket method, blunderingly catching in its blindfold grip both sick and well, the harmless and the harmful, indiscriminately. New-fashioned quarantine requires definite detailed knowledge applied with care and patience, not mere force'.[44] Sociologist Deborah Lupton writes similarly of this shift to government of the healthy self, the creation of a citizen/subject's expectations and desire for health:

> [P]ublic health and health promotion may be conceptualized as governmental apparatuses ... it is not the ways in which such discourses and practices seek overtly to constrain individuals' freedom of action that are the most interesting and important to examine, but the ways in which they invite individuals voluntarily to conform to their objectives, to discipline themselves.[45]

Lupton has also suggested that a major part of what characterised the emergence of the 'governmental' approach to public health in modern Western nations was the making of an explicit distinction between a new preventive model and the 'medieval' model of emergency response to containing disease – the emergency response of compulsory isolation.[46] All these sociologists think historically: the 'new public health' replaced or emerged out of something older and different. My object is to historicise some of the claims, precepts and assumptions of this literature.

Incorporating colonial governance into this historical-sociological interest in the development of the new public health in advanced liberalism, both demands and allows for a more complicated picture to emerge. Partly but not wholly because of the mapping of racial segregation onto health segregation, the practices and places of coercion and public health detention have been a stronger strand in this history and sociology of public health and governance than is often recognised. In Chapters 3 and 4, for instance, I compare the management of the chronic diseases of leprosy and tuberculosis. In a more or less contemporaneous way, new spaces of isolation were established for each: the leper colony and the sanatorium. Yet, in the colonial Australian instance, the nature, ideal purpose and effect of the enclosure of infected people could not have been more different. Almost exclusively white people were invited to voluntarily isolate themselves in the sanatorium, to enter into a six-month programme of 'open-air' treatment with the aim of re-training oneself into healthy and civically responsible conduct. There, their souls as well as their bodies were intensely governed. At the same time, white as well as Aboriginal, Chinese and Islander people were forced onto island leper colonies from the 1890s until the 1950s. There, their bodies, souls, conduct and civic identities were minimally governed. This was one example of exile-enclosure in the late modern world, which needs accounting for not just as an exception or a residual practice of an old preventive model, but as a deliberate innovation of the period. In the Australasian colonies and across the Empire and the twentieth-century Commonwealth, public health segregation and detention *said things* about governance and citizenship.

Imperial hygiene

In exploring government, hygiene and its reach, the book moves conceptually and structurally from smaller to greater instances of public health borders and spaces: from examination of intimate bodily boundaries, to the dissection of urban spaces in epidemic times, to the implicatedness of health and hygiene in national borders, to imperially and globally regulated boundaries. I begin with a chapter on vaccination in the nineteenth century. This is slightly perverse: if the book is about separation of the pure from the infected, vaccination with cowpox was the deliberate infection *of* the 'pure'. But I begin with this precisely because debate about vaccination constantly invoked problems of purity and impurity, foreign bodies and boundaries crossed. I also examine the imperial and global travel of cowpox, a contagious disease like smallpox itself, in a world where contagion, like colonisation, implied contact. In Chapter 2 on a smallpox epidemic in Sydney in 1881, I discuss the emergence of a bureaucracy of health, the segregations of the diseased and the clean in quarantine zones, and the question of compulsory public health detention made highly problematic – alongside compulsory vaccination – in this carceral space of public health. In Chapter 3, I look at a different kind of public health space at the urban level – the sanatorium for consumptives in the early twentieth century. A hybrid disciplinary institution – part health-resort, part workhouse, part hospital, part school – one aim of the sanatorium was instruction into safe and responsible citizenship. In Chapter 4 I locate leprosy as a new problem of 'imperial hygiene', one connected to the spate of race-based exclusion acts in the 1880s and '90s, which I examine in Chapter 6. Segregation and leprosy was widely discussed imperially and internationally, and the Australian policies were notoriously amongst the most rigid in the world. In these chapters, then, I begin to develop the connection between health, liberal governance and civic identity through specific substantive issues and events.

In Chapter 5, I move from urban and island spaces and enclosures to examine the national boundary of the new island-nation of Australia, which became a maritime quarantine line. The enforcement of this quarantine successfully kept cholera amongst other diseases, out of the island. But it also functioned centrally in the imagining of Australia as 'virgin' and uncontaminated, as clean and uninfected: the quarantine line was also a racialised immigration restriction line. In Chapter 6, I locate the various exclusion acts of white Australia within an international context. Such exclusions, deeply related to public health, were in fact internationally more normal than exceptional for the period. They were part of what came to be called 'international hygiene', new lines of communication and travel across the globe crossed by new legislative barriers of race-based medico-legal border control which were instituted in many nations from the 1880s. In the twentieth century, medico-legal border control became increasingly eugenic, screening out not only people 'undesirable' because of their race, but whites prohibited for other 'undesirable' characteristics as well. In the final chapter, I discuss sex both as contagion and as reproduction. What I call eugenic *cordons sanitaires* came into play as public health experts and eugenicists became increasingly concerned with the future population. All kinds of eugenic and public health mechanisms can be understood as *cordons sanitaires* between current and future generations.

All these were the lines of hygiene, the boundaries of rule which, in the modern period, formed populations and separated them, which instilled self-governance in some people and marked others as incapable, which excluded certain populations from certain spaces, which delineated islands as nations, people as races, and nations as pure. This is part of the critical history of colonialism, nationalism and public health.

1
Vaccination: Foreign Bodies, Contagion and Colonialism

It is well recognised that epidemics of communicable disease have long been a ramification of contact between cultures and communities, accompaniments to exploration, migration and colonisation, one of the events of the 'frontier'. Although what constituted the matter and the mode of contagion was constantly under dispute through the centuries of modern colonialism,[1] it was clearly recognised at the time and in subsequent scholarship that indigenous populations in particular succumbed to any number of fevers and poxes.[2] And, although the result of vastly different configurations of power, it was also recognised that the British – migrants, military, missionaries – suffered and died from 'alien' diseases in alien places,[3] hence as I discuss in later chapters, the discipline and institutions of tropical medicine. A concomitant interest in isolated individuals and communities, their vulnerabilities and immunities, has accompanied colonial and global epidemiology and public health, from nineteenth-century studies of 'pure' native tribes to the 'isolates' of the 1960s International Biological Programme.[4] Indeed, as we shall see through the book, public health administrators sometimes advocated enclosed segregation on public health grounds, in a way which mimicked this idea of natural isolates: indigenous people whose vulnerability sometimes justified a kind of health-based protective custody. Nineteenth-century public health experts were fascinated by natural isolations, commonly describing such communities as 'virgin soil', part of the dominant 'seed and soil' metaphor for understanding contagious disease.[5] Epidemiological, biological and racial ideas about im/purities and vulnerabilities of certain populations have historically mapped into military-colonial discourses of strength, defence, and resistance. Contagious disease in all these respects is clearly part of the history and historiography of colonialism.

In this history of contact, contagion and colonialism, smallpox has been extremely significant for several reasons. Its immediacy, visibility and virulence made the disease-ramification of contact between infected populations and previously uncontaminated populations clearly evident at the time, and intriguing to epidemiologists and medical historians ever since. Smallpox was a depopulating disease for many indigenous communities. This was the case in North America and Australia especially.[6] Moreover, smallpox is significant because for many aetiologists and public health practitioners in the nineteenth century, it was the model contagious disease.[7] At the same time it was different from other contagious diseases because its main method of prevention came to be not sanitary measures or isolation (although these were in place),[8] but rather vaccination. Smallpox was the first communicable disease to be prevented thus, initially through the practice of inoculation with actual smallpox matter (variolation) and from the very late eighteenth century through the practice of infecting preventively with cowpox matter (vaccination). What I explore in this chapter is an unfamiliar angle of this history of smallpox, contagion and colonisation: the idea that vaccination itself also has a history and a geography strongly associated with colonialism and settlement, with movement and contact. If infectious or contagious disease circulated between people and populations across the colonial globe, so did this first vaccine, the smallpox vaccine, both as actual matter and as technology. In this first chapter then, I think about vaccination as a kind of colonial contagion, as the deliberate circulation and proliferation of contagious matter along the imperial lines, and across the colonial borders of trade, travel and migration.

If this book is about social and physical borders, and hygienic practices across both, I begin with the skin-border, the membrane, that which the process of vaccination necessarily pierced and broke. This was no mere 'metaphor of invasion' – a Victorian anxiety amply studied and to which I certainly return[9] – but an all too real cutaneous introduction of a foreign-body into the self. The prevention of smallpox through cowpox also drew into consideration and management other social borders, including the classic urban and maritime *cordons sanitaires*. Separating the diseased from the clean was, by the nineteenth century, a longstanding response to illnesses comprehended as contagious, passed on by contact: the 'plague town' a dreaded but familiar idea; shipping quarantine measures and places were well established; the European cholera epidemics of the early nineteenth century were met by various measures of spatial policing. As I discuss throughout the book, versions of the *cordon sanitaire* proliferated in the nineteenth and twentieth centuries. But the prevention of smallpox was curious and in many ways atypical when viewed within this longstanding mode of prevention. Vaccination, by the late nineteenth century the primary preventive measure against smallpox, did not break the circulation of contagious matter in the classic mode of the *cordon sanitaire*. Far from separating out clean and dirty, vaccination rather involved the deliberate introduction of a diseased foreign body – cowpox lymph or dried crusts – into the individual and sometimes into hitherto uninfected 'virgin' populations. Thus in the history of the management of contact, and of the imagination and implementation of lines of hygiene between clean and unclean, vaccination offers immediate and intriguing complications. Vaccination crossed and dissolved the boundary between the clean and the diseased in an altogether different logic to segregation and quarantine. Moreover, in the colonial context, the 'foreign-ness' of these foreign bodies was not only a biological reference but often a racial reference, as 'lymph' (as the vaccine matter was called) circulated through many populations of children, literally linking them across the globe.

Vaccination has been comprehended and represented (in epidemiology, in medical history, in contemporary commentary) as being about movement, travel, contact and the circulation of infected bodies as well as vaccine matter and know-how from one part of the globe to another, from 'east' to 'west' and back again.[10] I explore some of these connections and circulations, drawing attention to the geographical and temporal tracing – the genealogical imperative – which characterised the procedure of vaccination over the nineteenth century, and into the twentieth. While there is an enormous literature on vaccination (and inoculation) I analyse it as not only contemporaneous with, but also as effected through and affected by, nineteenth-century colonialism. Vaccination, like contagious disease itself, was part of the connection of Empire.

Foreign bodies: boundaries and the logic of vaccination

Vaccination against smallpox had a precedent in inoculation. Inoculation involved the introduction of actual smallpox matter which was understood to cause a minor illness in the child, and thus the 'natural' smallpox or 'variola' infection would be prevented. Indian practices of inoculation drew much British scholarly attention in the eighteenth century and, famously, the practice within the Ottoman

Empire was introduced to England after Lady Mary Wortley Montague's experiences living in Constantinople in the early 1700s.[11] Vaccination by contrast, as Edward Jenner understood it from the late eighteenth century, introduced a different matter and its disease altogether. Lymph from a cowpox vesicle (on a calf or a human) was introduced into the body in order to set up the illness of cowpox, to prevent another illness – smallpox.[12] The point of cowpox was precisely that it was contagious; that it could be passed between humans by direct or indirect contact, and thus circulated through the social body. It involved the transfer and proliferation of vaccine matter through the social body as a contagion, the incorporation of one person's (or animal's) body into another's. Cowpox, or 'vaccinia' was a contagious disease in and of itself, and as such, involved a connection between humans; the direct contact of arm-to-arm vaccination, which was the favoured, if not the only method of vaccination through most of the nineteenth century. A group of children (and sometimes adults) would be vaccinated and were to return to the vaccinator usually on the seventh or eighth day, firstly so that the local reaction could be measured and assessed as successful or unsuccessful, and secondly in order that one or more of the children could be chosen to perpetuate the vaccine. A new batch of children would be infected with lymph from the arm of the original child or children. That group would return a week later, and another arm-to-arm process would take place. Thus, it was intended that the vaccine would be kept in circulation in an exponential way. While the use of stored lymph from calves became more common at the very end of the nineteenth century, most experts in the mid to late nineteenth century thought that the arm-to-arm method was preferable, that this process kept the lymph 'alive' and effective in a population of children.[13] As I discuss below, just who made up the population of children through whom the lymph had passed was problematised, sometimes in classed and oftentimes in racialised ways, in an increasingly colonial and global field of both smallpox distribution and vaccine distribution.

Immediately upon the expert endorsement of vaccination in England in the very early nineteenth century, and especially as the Imperial British government as well as some colonial governments flagged possibilities of compulsion for the procedure from 1853,[14] anti-vaccinationists proclaimed about the vaccine matter itself being a contaminant, and the procedure as contaminating: 'a filthy, disgusting animal poison'; a 'compulsory pollution of our veins'.[15] 'Vaccine lymph,' argued one pamphleteer – one of many hundreds – 'is an animal septic poison, and should therefore of course never be introduced into the blood of man, woman, child, or beast'.[16] Such comments, however rhetorically presented, were also a matter of fact: cowpox *was* a contagious disease, and (as we would understand it now) immunity *was* achieved through a process of infection. This transmissible aspect of vaccination was agreed upon by both opponents *and* proponents of the procedure. The contagious mechanism was not only decried by anti-vaccinationists but was the way in which those who supported vaccination theorised its effect and efficacy in individuals and later in populations.

Donna Haraway has written of the immune system that it 'is a map drawn to guide recognition and misrecognition of self and other in the dialectics of Western politics. That is, the immune system is a plan for meaningful action to construct and maintain the boundaries for what may count as self and other in the crucial realms of the normal and the pathological'. Haraway is right to call the immune system 'pre-eminently a twentieth century object',[17] for the mechanism of 'immunity' only existed as a tentative and disputed concept in the nineteenth century. While we are now familiar with the idea of immunity involving an *internal* self/other (mis)recognition (of foreign bodies by anti-bodies),[18] this was not readily available as a concept in the nineteenth century. *External* self/other recognition was, however. That is to say, vaccination involved the introduction of foreign/'other' matter into the integrated self/body. Just how vaccination resulted in immunity from smallpox was speculated upon, although most nineteenth-century practitioners were less concerned with physiological explanation than with empirically observable effect.[19] Sometimes mid to late nineteenth century physicians understood immunity as a local reaction, not a systemic one, as this New South Wales doctor did in 1881: 'I believe it is a local inflammation ... it is a concentration; any humours there may be are drawn to the vesicle'.[20] More often it was understood as a systemic process involving ideas about blood, and sometimes ideas about a lymphatic system. An 1875 rendition tentatively suggested the following actions: 'Vaccination ... have [sic] so impoverished some portion of the blood ... the healthy action of the lymphatic vessels, and glands (particularly of the mesentery), that when Small-Pox follows vaccination, there is not the same amount of pabulum, or food, for the poison to act upon and consequently less poisonous matter to be excreted by the skin'.[21] In 1883 Metchnikoff produced a new theory of immunity, one involving an active defence mechanism of the host and the principle of host resistance in the action of the phagocyte. Although the other great scientific figures of the time Virchow and Pasteur supported this theory, Koch opposed it, and in general, argues the historian of immunology Alfred Tauber, 'Metchnikoff's thinking places him outside the thrust of nineteenth-century conceptions ... [he was] misunderstood or ignored'. The more successful theory at the end of the nineteenth century (but not into the twentieth) was Pasteur's: 'the invading organism exhausted an essential nutrient during the first infection and was thus unable to survive in a host depleted of the substance. Such passive theories were the model of immunity and rested upon an ancient metaphysical understanding of health and disease, the balance of humours and the organism's ability to restore its wholeness'.[22]

What is immediately evident in Victorian discussion of vaccination-as-contagion is its structuring by logics of identification or similarity, and difference or foreign-ness. This is why homeopathic physicians weighed into the debate on vaccination so strongly, offering one model for understanding the mechanism and effect of vaccination in beneficial terms. Many homeopaths were interested in vaccination because it seemed to work around the principle *similia similibus curantur*.[23] One wrote: 'Vaccination is a homoeopathic diseasing measure: one disease is given to prevent a like one – vaccinia to prevent variola ... for in vaccinating a person we are *diseasing* him, we communicate vaccinosis to him'.[24] Notwithstanding such possibilities for understanding vaccination-as-contagion positively, it is clear that anxiety about this contagious and foreign quality to the vaccine was voiced in anti-vaccinationism throughout the nineteenth and well into the twentieth centuries: 'millions of people have now a ruined constitution through having the loathsome filth in the blood', wrote one.[25] In a culture where it was lay and expert commonsense that 'dirt ... is matter in the wrong place' as the physician Elizabeth Blackwell put it, well and truly prefiguring the anthropologist Mary Douglas,[26] this production of health through disease seemed counter-intuitive for many. The incredulous question was: 'Can disease protect health?'[27] Those who proposed and supported vaccination asked other practitioners and the public to understand that 'health' could be achieved through a process of infection and cross infection across multiple boundaries, including, possibly most problematically, species boundaries: the vaccine had an origin – whether in the recent or distant past – in animal disease.

The lymph from calves was drawn from cowpox pustules on the udder of the animal. One method of infecting the cow was by wrapping the animal in a blanket used by a person who had died of smallpox, and drawing the lymph from vesicles that formed on the udder.²⁸ The crossing and re-crossing of species and hygiene boundaries in such procedures was more than many nineteenth-century medical, lay and political sensibilities could tolerate or even imagine. Vaccine lymph, it was often argued, 'might engender diseases which you might never get rid of ... you have diseases of animals to consider, which are more to be dreaded than those proper to man'.²⁹ This inter-species exchange, the very idea of introducing a fragment of *diseased* animal into the human frame, was sometimes religiously, sometimes popularly, sometimes expertly opposed. At the very least, problematising the animal origin of the vaccine prompted arguments for maintaining the vaccine in the human population only, resisting the theory, discussed below, that the matter needed to be strengthened by sending it from the human population back through the cow's system. At most, a working-class British response to vaccination – as reported by anti-vaccinationists – was to describe animalistic features in recently vaccinated children. Thus mothers spoke of small horns growing in the heads of their infants, or that the voices of these children began to change to animalistic grunts, a response heavily satired. Vaccines were sometimes understood to produce unnatural hybrids: it was against nature.³⁰ For Hindus in India, on the other hand, the problem was one of the violation of the sacred animal itself.³¹

Because of the contagious nature of cowpox, it was also human diseases which gave cause for alarm, and for this reason calf lymph steadily replaced 'humanised' lymph in many countries, especially as governments increasingly regulated public health procedures toward the end of the century.³² The shift toward calf lymph was largely a response to well-placed anxieties about the transmission of syphilis and some other diseases between children in the arm-to-arm method. J.W. Beaney, a prominent Melbourne surgeon with expertise in venereology wrote in 1870:

> '*what practitioner is able to determine the purity of the lymph?* It is often far beyond his power to know the constitution of the parent of the child from whom the lymph is taken, or the nurse by whom it has been suckled; hence the difficulties that lie in his way are insurmountable, setting aside the ... latent germs that may lurk in the child ready to be transmitted through its lymph to others'.³³

While syphilis was the major concern, the possibility that vaccination was a conduit for other diseases and conditions was also raised.

Antivaccinationists published material on 'Cancer: a result of vaccination' and sometimes argued that leprosy was newly transmitted amongst Europeans in the tropical colonies because of the increase in smallpox vaccination.[34]

The discussion on the contagiousness and dangerousness of vaccination also crossed over into the question of inheritable conditions. The hydropathic physician John Marx, who was examined by an 1872 Committee on Vaccination in New South Wales attributed his own 'weak eyes' to vaccination; not his own, but his mother's: 'My mother was vaccinated, and shortly after the vaccination the glands of the neck swelled up, and the disease flew to her eyes. She was bad until a few months after she married. I was her first-born, and I inherited the complain ... I attribute my complaint indirectly to vaccination'.[35] His statements illustrate the intriguing and important nineteenth century conflation of heredity and infection, the ways in which certain diseases and conditions were understood to transmit both across populations in the present and between generations in time.[36] This complicated and connected history of concepts of infection and of hereditable transmission was especially important to the interwar eugenic refiguring of social and psychological conditions and 'tendencies' as contagious between generations, as I discuss in Chapter 7. But early expressions of the discourse of degeneration which came to hold such currency, are evident in these mid-Victorian vaccination debates. 'If Vaccination is allowed to continue', wrote one doctor, 'and the germs of septic poison (however light) are introduced into the blood from one generation to another, I think the result will prove to be most disastrous ... it has already been so, as seen by the pale and unhealthy appearance, of the rising generation, who are in no way fitted to stand hardship, or resist disease'.[37] This, it was argued, was the cumulative effect of vaccination over generations.

Essentially a diseasing procedure, vaccination did not fit nineteenth-century public health strategies at all easily. It did not fit the 'sanitising' trajectory of public health that was strongly comprehended through moralising 'improvement' arguments, through bourgeois imperatives of cleanliness and ordering. Medical historian Christopher Hamlin has suggested that nineteenth century public health was about the production of disciplined behaviour and new kinds of citizens drawing from 'a great confidence that people could change (or be changed) for the better'.[38] Within domestic British class and gender politics of public health, vaccination was simply not imaginable in these improving terms. There was very little possibility of attaching a moral purpose to vaccination, of imagining it as necessary for the shaping of conduct, or for the reform of domestic and social spaces.[39] Nor could it be attached easily to reformist activism, to the emerging social sciences, or to the crucial Victorian philanthropic networks and practices, all of which implemented sanitary reform and public health in the nineteenth century in one way or another as an improving and cleansing mission: vaccination could hardly be promoted as good for the soul. Rather, vaccination was a reductive, functional and biological procedure which, moreover, involved very little contact with public health experts or sanitary personnel. Indeed, insofar as vaccination *was* comprehensible within a moral/religious frame, it was to argue against it as an unnatural, irreligious and polluting practice. The argument *for* vaccination usually employed statistics as the authorising knowledge. But statistical argument only went so far in a period when public health was firmly determined by improving sensibilities, either in the political economy tradition or the philanthropic tradition.

If vaccination did not fit into the sanitising and improving tradition of nineteenth century public health, it also fell obliquely into the trajectory of public health isolation or segregation. There was certainly some European spatial isolation of smallpox sufferers. This was the case in London for example, during the epidemic of 1871–72 (in the light of the 1866 Sanitary Act).[40] In cities of global migration and sites of the Chinese diaspora such as Sydney or San Francisco, places where smallpox was not endemic, isolation was a more common practice.[41] But the availability of vaccination, which most Victorian governments met with some enthusiasm much as they debated the question of compulsion endlessly, complicated the idea of isolation considerably as I discuss in the next chapter. Indeed preventive measures which *did* function on the segregation and/or cleansing hygienic model were often set in deep and clear opposition to vaccination. For vaccination as a polluting procedure ran counter to the segregating logic of quarantine, of the clear imposition of boundaries between the clean and the dirty. This drove, for example, the fascinating anti-vaccination activity in Leicester, where notification and isolation policies were practiced in defiant and explicit opposition to compulsory vaccination laws.[42] Similarly, given the epidemic not endemic status of smallpox in the Australian colonies, it was sometimes suggested that smallpox could and should be controlled solely by maritime quarantine and compulsory isolation, that there should be no vaccination in the colonies, and that Australia could thus be both a cowpox and a smallpox-free island.[43]

The procedure of vaccination enjoined experts, governments and lay people alike to subscribe to the counter-intuitive idea of deliberately introducing particles of disease into otherwise healthy (and they were required to be healthy) human bodies. Boundaries between bodies were thus dissolved, and boundaries between species were crossed and recrossed. The process of vaccination manifested as a troubling confusion of the normal and the pathological as a minute amount of a pathological foreign body became a normalised part of the self, and a bit of one's own transformed body became part of another. This dissolution was not a comfortable one in a culture otherwise anxious to secure clean, firm and stable boundaries.

Connections: empire and the genealogies of vaccine

Despite the longstanding endemic status of smallpox in Europe, its origins have been consistently sought and asserted elsewhere by medical and government experts, and latterly by medical historians and microbiologists. Typically, the origin of smallpox was and is offered as generically 'Eastern', sometimes meaning the 'Orient', sometimes meaning the 'Far East'. This is the case in both nineteenth and twentieth century representations. For example the British educated physician L.H.J. Maclean told the New South Wales parliament in 1881: 'I believe it is an endemic disease of the valley of the Yang Tze and other river valleys in China ... It never made its appearance in Europe until about 1,000 years ago. It was only about that time that communication with the extreme east of Asia became fairly common, and it was with the commencement of that communication that small-pox made its appearance and small-pox came from the East'.[44] To take another example from the end of the period under study in this book, a 1954 *Story of Medicine* states that 'smallpox was first introduced into Europe by the Crusaders who brought it home from the Holy Land where the disease was common'.[45] And from a recent article by the distinguished scientist Frank Fenner: 'Speculatively ... variola virus ... evolved from an orthopoxvirus of animals of the central African rain forests ... some thousands of years ago, and first became established as a virus specific for human beings in the dense populations of the Nile valley perhaps five thousand years ago'.[46] Both the genre of medical history and the deeply related field of epidemiology seek origins almost pathologically. They are often driven by the conventional narrative imperative of a 'beginning' or of a 'case-one'. This historical and epidemiological origin-seeking has structured both simplistic and sophisticated accounts of the history of smallpox, as of other diseases which I discuss through the book. But origin-sourcing often serves hegemonic interests in the global, representational, scientific, national and racial politics of epidemic and transmission.[47] In the next chapter for example I discuss the pressing need for New South Wales politicians and medical experts to determine the presence or absence of smallpox in the 'virgin' Australian continent prior to 1788, that is, prior to English settlement. Yet asserting often mythical (and predictable) global origins of microbes is less significant than careful tracing of the more recent and verifiable effects of contact and contagion. The history of many colonial projects have been marked by the tragic path of diseases as they followed exploration, trading and transport routes.[48] The convention of global and colonial tracking of disease – in this case smallpox – can usefully be complicated through analysis not only of the microbes of the disease itself, but the microbes of the preventive disease, cowpox. Like smallpox, cowpox was itself spread through the colonial world along routes of human transport, travel, and communication. Thus, for example, a lone fur trader vaccinated local Cree people in the Canadian plains in 1838, with vaccine supplied by the Hudson Bay Company in the wake of an already devastating set of epidemics.[49]

As technology and as matter, vaccination was part of what connected Empire. This was so from its earliest introduction. One account of the reception of vaccine in Ceylon from the first decade of the nineteenth century reveals the precarious communication of the matter over land and sea from London.

> In the course of the last twelve months, we have repeatedly received by sea from England, the Vaccinate Matter, and many children have been Inoculated to no purpose ... Fortunately, Dr Short, a surgeon of this Establishment residing at Bagdad produced the disease at that place. He immediately forwarded the matter to Bussorah ... Mr Milne soon afterwards inoculated a number of other Children and he sent the Vaccine Matter to Bombay, by several ships.

The author of the account indicated that 'we now have it in our power to communicate the benefit of this important discovery, to every part of India, perhaps to China, and the whole Eastern World'.[50] Like the communicable disease of smallpox itself, then, vaccine matter moved around the globe, sometimes through populations, sometimes as lymph or crusts stored in vials, or as saturated pieces of linen.[51]

There were certain centres for source production – initially Jenner's National Vaccine Establishment in London – which sent vaccine off to 'colonise' new regions with missionaries, traders, bureaucrats and militaries. By the late nineteenth century, and because of the geography of Empire in which London might be the metropolitan centre, but might also be the furthest away (for example from the Australasian and Pacific colonies), sources for lymph were established in other colonies. In the 1880s the New South Wales government received regular despatches of calf-lymph from Bombay. It also sought and received from the government in Bombay detailed information and instruction about the methods of obtaining lymph from the calf (see Figures 1.1 and 1.2). These vaccine travels over borders, through Empire and beyond, occurred both temporally and geographically. Tracing a genealogy back in time and through known places and populations to an original source was consistently important, if disputed in major ways through the nineteenth century. There were two kinds of genealogies at issue which draw attention to the imperial and global politics of smallpox dissemination and prevention. First, the genealogy of the actual vaccine matter came to be understood as vitally important: vaccines had pedigrees, their own blood-lines. Second, and related, the genealogy of the children through whom the matter had passed was investigated, monitored and documented.

First, then, the genealogy, the 'lymphline' of the vaccine. On the face of it, inoculation involved the use of smallpox matter: vaccination involved the use of cowpox matter. But the apparently new technology of vaccination was not that simple. In some cases the vaccine derived from the pustules of the cow infected with cowpox. In other cases the vaccine originated from the pustules of a cow infected with smallpox, infected deliberately from a child with the actual smallpox disease. Those who supported this practice theorised that the disease turned into cowpox when it was thus 'bovinised'. Sometimes doctors argued that the most effective vaccine was that which had been in circulation in the human population for the longest time – 'humanised lymph'.[52] At other times doctors claimed that this weakened the vaccine. In these cases it was suggested that the vaccine matter needed to pass back through the cow periodically, in order to maintain its strength and potency, and thus 'bovinised lymph' was considered most effective.

Whether a physician or public vaccinator considered bovinised or humanised lymph to be more potent, the specific vaccine which any one practitioner used needed to be traced and traceable temporally and geographically – its 'line' known and verified. In England, this involved

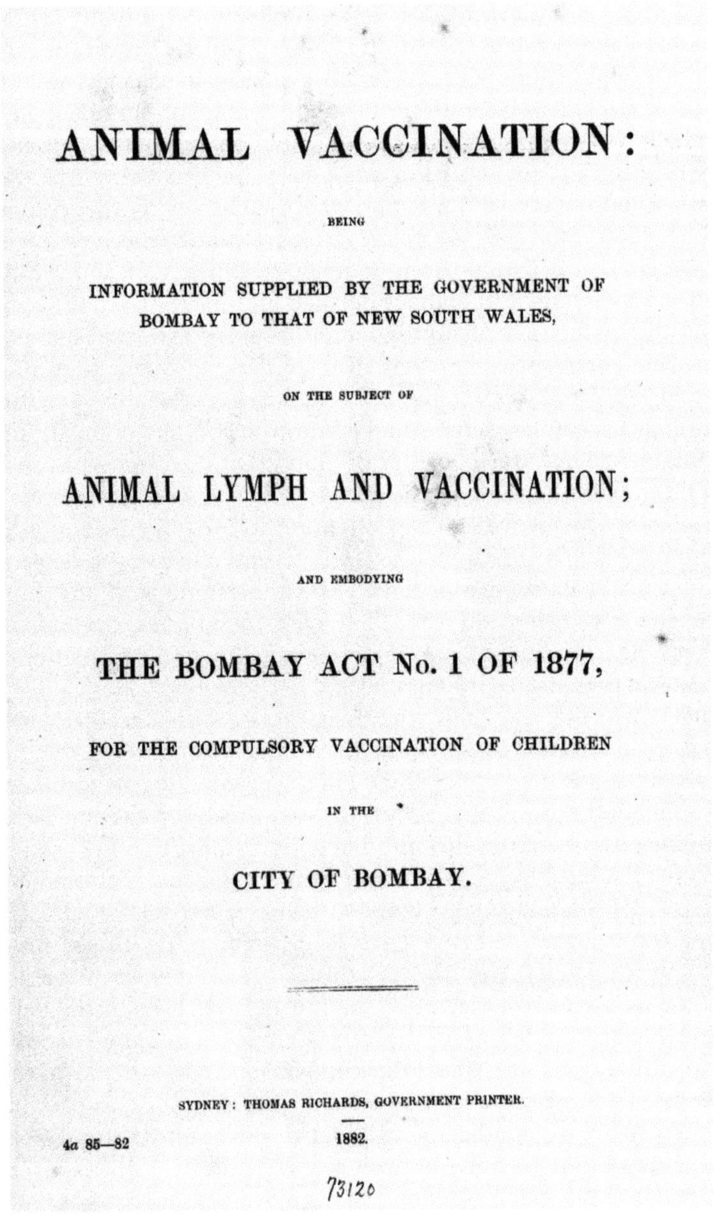

ANIMAL VACCINATION:

BEING

INFORMATION SUPPLIED BY THE GOVERNMENT OF
BOMBAY TO THAT OF NEW SOUTH WALES,

ON THE SUBJECT OF

ANIMAL LYMPH AND VACCINATION;

AND EMBODYING

THE BOMBAY ACT No. 1 OF 1877,

FOR THE COMPULSORY VACCINATION OF CHILDREN

IN THE

CITY OF BOMBAY.

SYDNEY: THOMAS RICHARDS, GOVERNMENT PRINTER.

1882.

4a 85—82

73120

Figure 1.1 Technology, as well as lymph, travelled between colonial governments
Source: *Animal Vaccination*, Thomas Richards, Sydney, 1882, p. 2. Courtesy Mitchell Library, State Library of New South Wales.

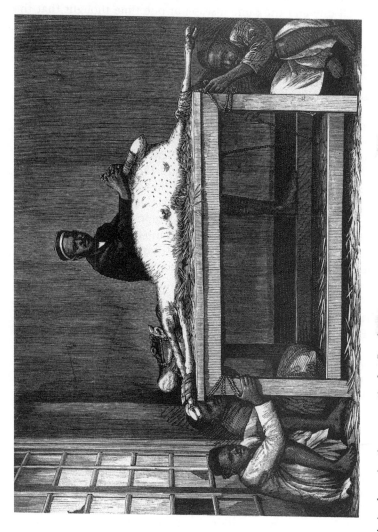

Figure 1.2 Animal vaccination as practised at Bombay
Source: *Animal Vaccination*, Thomas Richards, Sydney, 1882, p. 2. Courtesy Mitchell Library, State Library of New South Wales.

debating the origin of Jenner's lymph in terms of species and disease. One rendition held that

> Jenner believed that swine-pox, cow-pox, small-pox and the grease were all one and the same disease; at one time thought that the grease (the original source of the true vaccine) should be modified by passing through the cow, before being used as a prophylactic, but upon further enquiry he arrived at a different conclusion, and used the grease in its natural state and supplied the public with it from its original source, the horse's heels.[53]

In colonial contexts, this anxiety about the genealogy of the lymph itself was compounded by concerns to interrogate the global travel of the vaccine matter, since that origin. In New South Wales, at committees of inquiry on vaccination in 1872 and 1881, the 'purity' and the pedigree of vaccine matter in that part of the Empire came under constant question: could the vaccine matter be traced to Jenner's lymph? And if not Jenner's, then whose? Did the lymph originate in human or animal? And, especially, through which humans had it passed on its way from England to the Colony? One witness at the 1872 Committee detailed the possible lines of origin of vaccine matter available in New South Wales:

> I have been in the habit of getting vaccine lymph from Mr Badcock, in England ... This lymph is transmitted through the cow every six or twelve months ... There are three ways in which the lymph now used in England has been obtained. 1st The original Jenner lymph. 2nd Lymph obtained by Mr Ceiley who variolated the cow first in 1839, and from which course Mr Badcock's supplies have been obtained; and 3rd vaccine brought to England by Dr Blanc from the Continent, obtained from a cow under the natural disease, and reproduced upon heifers.[54]

He said that the lymph he received from England every second mail was 'pure from Home'. It was the 'Home-ness' in this statement which advertised this vaccinator's lymph as safe and clean in a context where there was a certain competition for purity amongst practitioners. Lymph direct from 'Home' implied that it had not passed through other colonial populations – that it had not passed through populations of Indian children in particular, given that southern India and Ceylon were necessarily ports of call for ships from England to the

Australasian colonies. Alternatively, 'purity' was announced because the vaccine was straight from the cow and had passed through no children at all, or very few. For example, advertisements for vaccination during the 1881 epidemic of smallpox in Sydney invariably announced the source as 'pure lymph from heifers', or 'Pure Vaccinate from the Heifer'.[55] Under constant discussion was the relative efficacy as well as purity of lymph direct from the cow, as opposed to 'that which has passed through many hundreds of constitutions'.[56] But even lymph pure from 'Home' or pure from the calf was not automatically established as of clean lineage, nor were anxieties about origins thus precluded. Even the purity, or more accurately the authenticity of Jenner's vaccine was under question, both in England and the colonies: 'the original vaccine lymph we use from the Royal Vaccine Institute has never been vaccine lymph at all, from the day it was introduced in London to the present time – it was derived from the arm of Jane King, who was inoculated from a horse with greasy heels'.[57] The 'True Pedigree of English Vaccine' as one English physician titled his chapter, was debated regularly, and in fact has been ever since.[58]

Nineteenth-century concerns about the vaccine matter were not just medical anxieties about biological purity, but were also cultural anxieties about race, class and species mixing. Given the standard practice of arm-to-arm techniques, the purity of the vaccine drew into consideration the purity of the individual child or adult, and the population through which the vaccine passed, and which necessarily became a literal part of the vaccine matter itself. The genealogy of the vaccine, then, was also a genealogy of the children through whom it had passed. This was often detailed with great precision, especially at the beginning of the nineteenth century when in some colonies very few children had been vaccinated. As I have discussed, some medical men argued for the increasing strength of vaccine the longer it stayed in the human population. But that implied contact with increasing numbers of unknown individuals, it implied connection – incorporation – with wholly unknown populations: had the vaccine matter been through a population of Indian children? Had it been through a population of children in east-end London or inner-city Sydney? Did it have any point of connection with groups compulsorily vaccinated at one time or another, one place or another – Chinese indentured labourers, nurses in the new infectious disease hospitals, criminals in gaols? Officials, doctors, public vaccinators, especially in the antipodean colonies where the temporal and geographical distance from the source was greater and anxieties often more acute, desperately sought and sometimes fabricated origins and paths through populations for their particular vaccine lymph.

It is also the case, however, that anxieties about the race of vaccinated and vaccinifer children were less evident in the first decades after the technique was introduced in England and the colonies and belong much more to the mid and late nineteenth century, and certainly to the first half of the twentieth century. Indeed tracing anxiety about vaccine connections is one way of tracing changing conceptualisations of racial difference and its significance. Distinctions between people changed broadly from eighteenth century classifications – fluid, alterable, based on a common conception of Man – to nineteenth century 'races' – irreducible, essentially unalterable, biologically different.[59] Conceptions of race, and conceptions of healthy and unhealthy racial populations were connected developments. As Mark Harrison has suggested, in the seventeenth and eighteenth centuries there was a 'virtual absence of fixed racial identities' in literature on India, but after 1800 'racial identities came to be fixed and ... India was viewed ... as a reservoir of filth and disease'.[60] Racial categories and the boundaries separating them refined and hardened over the nineteenth century, so that all kinds of connections and contact acceptable at the beginning of the century were unacceptable by the end. This has been discussed with respect to sexual contact between individuals defined and self-defined within the newly reified and biologised discourses of race. Sexual relations officially permissible in the eighteenth century became less so as the nineteenth century progressed.[61] The very different connection between vaccinated children, which was nonetheless also a literal biological mixing, is another site where this hardening of categories can be observed, where the biologising of race played out.

Government returns for vaccinated children in early nineteenth century Ceylon graphically respresent the child-to-child, arm-to-arm method which literally connected these children. One of the earliest returns from 1802 documented age, sex and caste ('European' or 'Pariah') in the manner of such returns throughout the nineteenth and twentieth centuries, even as the categories of identification changed (see Figure 1.3). But what is notable here is that 'Pariah' and 'European' children were vaccinated from each other, evidently without concern. At the same time, the source-child was still important to define and make known in the Jennerian tradition of genealogical tracing: 'A Fortunate inoculation at length produced the vaccine disease in Anna Disthill, who is perhaps the first human being who underwent it in India ... the daughter of a servant of Capt. Hardies ... It is necessary

Return of Patients Inoculated for the Vaccine Disease by A. B. Medical *****, the District of ***, in the month of September, 1802.

Numb.	Date of Inoculation	Patients' Names	Age	Sex	Cast	Place of Residence	From what Patient Inoculated	Nature and Period of the Fever	Nature and Number of the Pustules	When Discharged
1.	20th Aug.	Moutow	2 years	Male	Pariah	Trincomalie	John Sybille	Slight on the 27th	One distinct	2d September
2.	20th Aug.	Janet James	5 years	Female	European	Trincomalie	John Sybille	None observable	One distinct	2d September
3.	27th Aug.	Murga	6 months	Male	Pariah	Mullaivo	Moutow	Slight 3d Sept.	One distinct	9th September
4.	28th Aug.	Ram Sammy	28 years	Male	Pariah	Cotiar	Moutow	Severe 2d and 3d	Numerous variolous	20th Septemb.
5.	3d Sept.	Cotly	20 years	Male	Pariah	Trincomalie	Murga	Slight 10th Sept.	One distinct	15th Septemb.

TABLE SHEWING THE NUMBERS AND EVENTS.

Casts.	Number.	Discharged.	Remaining.
European	1	1	0
Pariah	4	4	0
Total	5	5	0

NOTE:—Moutow and Murga after having passed through the Cow-Pox were at the request of their parents, inoculated with variolous matter, which failed to produce any effect. The Disease in Ram Sammy was evidently variolous, and on enquiry it was found that previous to inoculation, he had been exposed to the contagion of Small-Pox, which is prevalent at Cotiar.

Figure 1.3 Return of Patients Inoculated, 1802
Source: Thomas Christie, *An Account of the Ravages Committed in Ceylon by Small-Pox*, J.&S. Griffith, Cheltenham, 1811. Courtesy, The Wellcome Library for the History and Understanding of Medicine, London.

to mention these circumstances as from her alone, the whole of the matter that is about to be sent all over India, was first derived.'[62] Officials certainly indicated that 'no pains have been spared to make it pass through unexceptionable bodies',[63] but in the first decade of the nineteenth century, at least in Ceylon, non-Europeanness itself did not make the child 'unexceptionable'. Later in the century, anywhere in the Empire, this was not the case, for by then biological purity and racial purity were deeply connected concepts.

By the late nineteenth century, part of what constituted an 'exceptionable body' in terms of vaccination, was not only race and class, but also 'syphilitic descent'. In New South Wales a child determined to be of syphilitic descent was discounted from the arm-to-arm lineage: in effect placed outside this social body.[64] The possibility of the transmission of syphilis from child to child was discussed regularly and with great concern. And so the genealogy of the child was also in question: the family history of the child, or more precisely the sexual history of the father. In one anti-vaccination pamphlet published in 1875, *Vaccination and its Evil Consequences*, the author asked: 'by taking lymph from one child and applying it to the arms of another, how do we know whether the father or mother, to say nothing of the grandfather or grandmothers, have not had Syphilis, Scrofula, Insanity ect [sic]?'[65] And in the 1872 Committee one doctor was asked: 'How can medical men guard against the use of impure lymph?' And he responded: 'The chief protection is in a knowledge of the family history of each child which in the city is difficult to get ... I think it would be wise to adopt a precaution they have in London, that lymph should not be taken from a first-born child in a family, because the first child would be the most likely to show a syphilitic tendency'.[66]

But when all was said and done, neither physicians, public vaccinators, mothers, nor governments ever really knew these lineages with any certainty. As the Sydney homeopath John le Gay Brereton said in 1872: 'You must know that medicine is not an exact science like mathematics or geometry, and you cannot trace a taint as you can trace a line. You can only draw an inference'.[67] Yet precisely because physicians, governments, and parents could not know conclusively, they sought obsessively the 'trace of the taint', the lineage of the vaccine. The genealogical lines of the vaccine and the lines of children through whom it passed and from whom it was derived were lines through the territory and population of Empire, from London to Ceylon, from Bombay to Sydney.

Vaccination and administration: certificates, scars and passports

I have suggested that vaccination did not fit easily with Victorian sanitary reform, or with Victorian modes of spatial management as a way of controlling disease. But there was a third aspect of sanitary reform and public health which the procedures surrounding vaccination both positively required and assiduously promoted – administration and bureaucracy, and their knowledges of statistics and demography. The medical and public health axes of statistics and administration were the techniques and rationalities which 'unified the inhabitants of geographical space as a social body ... through the charting of social and moral topographies of bodies and their relations with one another'.[68] Vaccination came to be important as a means for the collection of information through systems of registration and certification of individual infants and children. It provided one of the mechanisms through which British as well as colonial populations were rendered governable. Vaccination, like the tracking of epidemic disease itself, became part of the growing biopolitical business of population health, of collecting and producing the 'vital statistics' of the social body.[69] It helped build the vital statistics of Empire. Moreover, unlike the multifaceted sanitary projects of domestic order, personal hygiene, drainage and waste management, or even sanitary architecture, the simplicity and visibility of the vaccination procedure, its measurability, its easy transmission as 'fact' and thus ready morphing into statistical knowledge, meant that vaccination (and smallpox) rates, probabilities and projections were amongst the most statistically documented procedures in the nineteenth-century medico-administrative domain, both by British governments and colonial governments. And unlike other public health endeavours that remained in the philanthropic, voluntary, and reform sector (broadly) until early twentieth-century welfare, vaccination was a state-interested procedure almost from the beginning. In England and Wales vaccination was connected to the Anatomy Act, the New Poor Law and other shifts in working-class management from the 1830s.[70] In New South Wales which was strongly governed from its inception as a convict colony in 1788, the storing, production and administration of vaccine and vaccination, was very early the business of government.

Although the precise requirements changed over time and place, doctors and public vaccinators generally documented each individual vaccination and re-vaccination, information that was tabled and produced as government returns. Additionally, and significantly, systems of certificating vaccination were early implemented in many colonial contexts, as in England. The system was linked to, and provided further information for, the registration of births (and deaths) and the vaccination certificate, like the birth certificate was intended to be an essential item for each child and parent in an increasingly governed world. Importantly, duplicates of vaccination certificates were usually given to parents. This was required, for example by the Imperial Vaccination Act of 1867. This tied people into webs of governmental knowledge, made them known to local officers, registrars, practitioners, but also made certain kinds of statistical and epidemiological knowledge possible. In colonial contexts medical and sanitary interventions enabled a new kind of governmentalisation of the colonial state in which quantification, distribution and administration of an indigenous or a white 'settler' population was both achievable and produced new kinds of subjects for rule.[71] Vaccination created 'populations' in the bureaucratic sense: people were both individualised by the procedure, and aggregated. In contexts like India where information on birth and death was very difficult for the governing bodies to collate, vaccination returns provided considerable data.[72] The creation of an abstract population as data from vaccination returns offered a 'field of visibility' to government, what sociologist Mitchell Dean describes as a mode by which it was possible to ' "picture" who and what is to be governed, how relations of authority and obedience are constituted in space'.[73]

Yet vaccination programmes created populations not just abstractly but as actual incorporations as well. The delineation of 'vaccination districts' for example was not simply an arbitrary administrative act, but took its significance from the very literal problem of population within an area for the success or failure of a vaccination program. Because there was a limited time in which lymph could be taken from a cowpox pustule, the arm-to-arm technique required the procedure to be constantly in process: there was an in-built imperative to vaccinate many children within a geographical and numerical population. For example, after its precarious route through Baghdad, and because dried lymph sent directly from London had failed, the vaccine matter was precious in Ceylon. The Medical Superintendent in Ceylon directed local medical officers thus:

> You will consider yourself as entrusted with the care of keeping up the Vaccine Matter in your district, and of constantly preserving the Virus in a recent state, for which purpose you will be careful by suc-

cessive Inoculations ... to perpetuate the disease, on living subjects, so that you may at all times be provided with Patients, from whom recent Matter may be taken[74]

Later in the century, one argument for compulsory vaccination was the need for a population of a certain size in order to keep the vaccine 'alive' and in circulation. The medical adviser to the New South Wales Government in 1869 said that 'to keep up a good supply of vaccine virus, it is necessary that vaccine districts should not be too small; that a population of not less than 25,000 should form a vaccination district ... In such a district, with a Compulsory Act, it would be easy to keep up a full arm-to-arm vaccination'.[75] This suggests how the technology of vaccination made the concept of 'abstract space' of administration, as well as the abstract concept of 'population', into biological phenomena. That is, unlike other governmental systems for data collection or even sanitary or therapeutic intervention, the arm-to-arm method of vaccination constituted groups of people as actually incorporated populations: vaccination connected them, one to another in an exponentially increasing way. If many recent scholars of biopolitics have become interested in the means by which the juridico-political metaphor of the social body became a biological object and field for intervention, the instance of arm-to-arm vaccination offers itself quintessentially, in its crossing of metaphorical and abstract representation, its administrative implementation, and its always literal incorporation of bodies within populations.

Smallpox was the most visible of diseases, leaving its survivors permanently pock-marked and disfigured, wearing the stigmata of the ill. But if the sufferer did survive, their very scars then marked that person as immune. In this logic, having no mark at all, being completely 'pure' if you like, was far more suspect. This was the threat of an incubation period, what by the 1890s came to be configured as 'the carrier' – one may well be diseased but not yet show the signs. While smallpox was a contagious disease that rendered sufferers unmistakably marked and visible, there was nonetheless always the threat and insecurity of the incubation period, its moment of invisibility: unmarked people therefore were not necessarily 'clean'. Increasingly, for many governments, to have one mark, the single pock mark of vaccination, rendered the disease status of that person known, conferring an immunity to disease and an immunity to travel over governmental lines of hygiene. Yet because of the strange, 'foreign body' logic of vaccination, the scar did not signify cleanliness: in this schema, purity itself (neither having had the smallpox disease nor the vaccine disease) became danger. While medical historian Nadja Durbach has shown how the vaccine scar was understood by working-class English objectors as a kind of criminal brand for officials, the scar also signified safety and security, bodily evidence of the deliberate introduction of the foreign body of the cowpox vaccine.[76]

In times and places of emergency quarantine, either the vaccine scar or the scars of the disease itself conferred a capacity to move (more) freely across all kinds of borders. In the 1881 epidemic of smallpox in Sydney which I discuss in detail in the next chapter, there was a quarantining measure established at the border between the colonies of New South Wales and Victoria. Passengers on the Sydney–Melbourne train would be examined at the border town, and only those with certificates of vaccination and a viable vaccination scar were permitted entry into the 'clean' colony of Victoria.[77] In 1913 smallpox again appeared in Sydney and the new Commonwealth Government declared all of Sydney a quarantined space, with road blocks established in a circumference 15 miles from the City Post Office. Again, people were permitted in both directions over those barriers only if they could display a viable vaccination scar.[78] The vaccination scar thus facilitated movement into 'clean' spaces, but it could also get one safely into 'unclean' spaces (a quarantine station, an infectious disease hospital, a segregated street or building), over borders and over lines of hygiene.

In some contexts the scar and increasingly an accompanying vaccine certificate became quite literally passports into and out of certain zones. In many ways prefiguring, but at the very least accompanying the invention of the passport itself, about which John Torpey has written extensively,[79] the vaccine scar and its certificate were early documents of identification, travel and passage. The vaccine scar and/or its documentation granted an 'immunity' to travel over national borders. Over time, and especially in the newly bureaucratised culture of the early twentieth century, vaccination was to be recorded on the body as a scar that *needed* to be visible to be viable and recognised, but also recorded by an emergent health and immigration machine. In some cases a vaccination scar or certificate was an immigration requirement into certain nations: a passport and often part of compulsory official documentation even before passports themselves were commonly required. In places like the Australasian and Canadian colonies, where smallpox was epidemic, the vaccination passport became increasingly common, especially for those seeking immigration. From the 1870s, English emigrants to the Australian colonies were required to be vaccinated if small-pox was present in the English region from which they

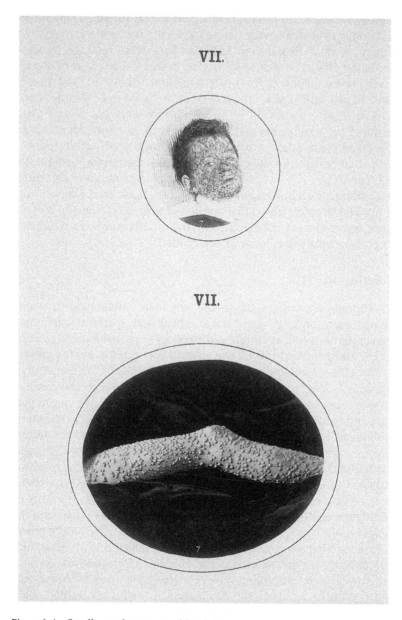

Figure 1.4 Smallpox: the most visible of diseases
Source: J. Ashburton Thompson, *A Report to the President of the Board of Health containing photographs of a person suffering from variola discreta, and accounts of the case*, Government Printer, Sydney, 1886. Copy in author's possession.

came. This was a condition of embarkation.[80] Similarly, documentary and bodily evidence of vaccination (or smallpox itself) was required of emigrants from England to Canada. They were required to show Canadian officials evidence 'of having been vaccinated, or of already having had smallpox'.[81] From 1908 a medical 'Inspection Card' needed to be signed, and kept for three years and 'shown to government officials whenever required'.[82] In Australia a 'Personal Detail Card' recorded the 'vaccine history' of an individual, not only if vaccination had taken place, but where, when, by whom, with what vaccine, and with what reaction (that is, size, colour, and discharge).[83] I return to the connections between vaccination, documentation, infectious disease and global movement throughout the book, but in these developing and increasingly global and governmental systems of surveillance and of identity documentation, the vaccine-scar was a significant corporeal identity document in and of itself.

Conclusion

The histories of transmissible diseases and their containment have long invited attention to geographical and historical, spatial and temporal axes of transmission, attention to the history of movement and contact, and to the history of imperial and colonial connections. But the preventive vaccine for smallpox also produced a transmissible disease. And its 'epidemic' spread through imperial and global individuals and communities was not incidental to but necessary for its success. The spread over time and space of the technologies of vaccination, of infected/immune individuals and populations, and of both the vaccinia and the variola viruses, implicates smallpox and vaccination in a modern history of travel and colonisation. The movement of the vaccinia virus in stored vials or cloth, or in the pustules of children's arms traced the global lines of Empire in the Victorian period. These were also the lines of knowledge, as the technique was disseminated with the matter and the disease itself. This is the colonial history of vaccination. But the procedure of vaccination *crossed* lines as well as travelling lines of communication. Alongside the question of compulsion, this crossing accounts for the extraordinary noise about the practice in the period – expert, religious, political, and popular. Vaccination crossed the membrane-line, introducing a foreign body into the otherwise healthy self. It crossed species lines. It enabled the crossing of governmental lines of hygiene – quarantine and segregation borders, and later national borders. Most importantly, it crossed the line between the pure and the impure.

2
Smallpox: The Spaces and Subjects of Public Health

In the modern period, places of infectious disease segregation were heterotopic: they could be governed simultaneously by the imperatives of a penal system and a health system, and as we shall see in later chapters, informal and sometimes formal systems of race management.[1] Penology and public health were both segregative discourses. In 1881 there was an epidemic of smallpox in Sydney, the management of which offers a fine historical example of the dovetailing of the penal and the medical. Such epidemics were not uncommon in the colonial period,[2] but this one is interesting because it gave rise to legislation that allowed for the forcible containment and segregation of people as a way of containing infectious disease. Modelled on the 1832 Act to prevent the spread of cholera in England,[3] the Infectious Disease Supervision Act New South Wales (1881) also prompted the establishment of a Board of Health in New South Wales; 'health' was thus bureaucratised in the self-governing British colony.[4] The Quarantine Station, where smallpox sufferers and their contacts were compulsorily detained, was a place of isolation for public health reasons, but the initial segregation and the *cordons sanitaires* themselves were implemented and maintained through policing and punishment measures. Quarantine was about health segregation but it was also a form of enforced detention, a carceral place. Therein lay the problem for an emerging liberal administration. The question of compulsion and liberalism has been considered at length by many historians with respect to vaccination. But with the exception of the British and colonial Contagious Diseases Act, and to some extent discussion of leprosy confinement, public health detention has been far less focused upon.[5] Many colonial governments in the late nineteenth century had to deal with the strong objections to segregation for health reasons: this was the case during the 1896 plague in India, for example.[6] Here, in a study of one particular epidemic and its responses I explore how detention and compulsion produced governance and consent as a problem. This was an expression of the paradox of liberal governance: both the heavy administration of the epidemic and questioning of compulsory segregation which was part of that administration, were formative sites in the genealogy of liberalism and of the liberal subject.

If in the previous chapter 1 explored the ways in which the procedure of vaccination against smallpox dissolved the hygienic boundaries separating 'clean' and 'infected/immune' bodies and populations, in this chapter I set out one government's late nineteenth-century attempt to draw rigid borders between the infected and the uninfected, the declaration of *cordons sanitaires*. If epidemic disease was about place, the longstanding response of administrative government was to create borders between clean and unclean spaces and to move people in and out of these spaces for certain periods of time: the technology of quarantine. In 1881, Sydney was divided by numerous geographical and administrative lines of hygiene. I focus in particular on the geography and uses of the Quarantine Station at North Head, the liminal place where infected people and their contacts were required to go. This quarantining strategy was certainly only one of several official responses,[7] but I have chosen to concentrate on it because mass quarantine of this sort was an unusual public health practice by this time, internationally speaking.[8] Moreover, in that smallpox was the most threatening quarantinable disease in the Australian colonies, there was a constant effort to integrate isolation and vaccination as preventive strategies. But events at the Quarantine Station illustrate the very different preventive logics of the two strategies, one separating the clean from the unclean, the other figuring the 'clean' or uninfected person, as suspicious and seeking to 'contaminate' the latter with cowpox vaccine in order to protect them.

Smallpox in Sydney, 1881

In the nineteenth century, public health bureaucracies were formed, expanded or strengthened largely in response to epidemics of communicable disease.[9] As Charles Rosenberg has written, 'accepting the existence of an epidemic implies – and in some sense demands – the creation of a framework within which its dismaying arbitrariness can be managed'.[10] Medical jurisdiction, public health and later 'state medicine' were expanded via the management of epidemics which required an apparatus of, and for, intervention. Cholera, the disease so formative of public health instrumentalities in North America and Britain, never arrived on the Australian continent.[11] Smallpox did however, and there were intermittent epidemics through the nineteenth and twentieth centuries, beginning with that which killed many Indigenous people in the Sydney area after the arrival of the first fleet of convicts and military personnel in 1788.[12] The epidemic I analyse here started almost a century later. It was recorded as beginning in May 1881 with a Chinese child who lived in George Street, in the middle of the city.[13] The house in which this family and their contacts lived was quarantined; a police guard was placed at the front of the house preventing entry or exit of any person, and monitoring any communication. After 21 days, several other cases appeared in the inner city. In each of these cases, the medical and police response was similarly to isolate individual houses and the people within them.[14] A programme of vaccination was initiated. After some months, and a relatively slow escalation of the number of infected people, it was decided to abandon the policy of quarantining individual houses in the city, and instead to remove infected people and their contacts to the Quarantine Station on the North Head of Sydney Harbour, which had housed infected people from 1828 and had more formally received 'infected' ships since 1837.[15] What became known as the Sanatory Camp was later established at Little Bay south of the city, where a smallpox hospital was built, and where many Aboriginal people had suffered and died from smallpox.[16] This was the same isolated coastal region where the tuberculosis sanatorium which I discuss in Chapter 3 was established, and where people with leprosy were later sent. By the time the smallpox epidemic was declared to be over, 41 people had died, and 163 people were recorded as infected. Many times this number were implicated as contacts, isolated in houses, taken to the Quarantine Station or to the Sanatory Camp. For example, when a Mr Rout was diagnosed with smallpox in June 1881, at least 14 other people in his household were also immediately isolated.[17] In 1882–3 there was a Royal Commission into the management of the epidemic and another specifically into the conditions and management of the Quarantine Station. Additionally, an enquiry was established to investigate (again) the question of compulsory vaccination in the colony.[18] One of the doctors involved, Edinburgh-educated government medical officer S. Mannington Caffyn, wrote that the event 'upset the business of the colony as much as a civil war could have done'.[19] This was not, as in other epidemics, because of the casualties: the number of people sick and dead was really not that high (and this indeed was seen to indicate the success of the swift and forceful quarantining measures). Rather it was because of the questions of governance, freedom, consent and compulsion raised and practiced so explosively in the management of the epidemic, and which concerned detention in the first instance, and vaccination secondarily.

The formalisation of a public health bureaucracy in New South Wales began, significantly, with this smallpox epidemic. The Infectious Disease Supervision Act, assented to in December 1881, required the appointment of a new Board of Health. The Board consisted of the mayor of Sydney, the Under-Secretary of Finance and Trade, the Inspector General of Police, six doctors and the Colonial Architect. The 1881 Act gave the Board of Health 'powers to isolate'. It also required cases of smallpox to be reported to 'proper authorities' which meant 'the nearest police-station or lock-up' or to 'the officer-in-charge at the Central Police Station'.[20] The bureaucratic history of public health in New South Wales points to the presence of forceful sovereign power at its moment of inception. The liberty of individuals was withheld for the larger benefit of safeguarding the public health, in this case, containing the spread of smallpox. Modelled on early nineteenth century English quarantine powers – outdated by 1881 – the most similar local measures were those derived from lunacy incarceration powers, also strongly involving the police in their implementation.[21]

On the other hand, however, the New South Wales Board of Health brought together what is a fine example of a 'biopolitical' constellation of interests. The presence of the bureaucratic representative of Finance and Trade implied the direct impact of quarantine and epidemic on commercial interests, and more generally, the significance of the health of the social body for the capitalist economy. Population health, especially with respect to infectious disease, was not only an internal matter, but constantly brought into consideration trading relations with other economic and governmental entities. Governmentality was also about security. The presence of police as agents of public health clearly indicates the proximity of a government desire for 'health' with a government desire for the maintenance of public order. The police in this instance functioned as diluted versions of the more totalising eighteenth century German medical police,[22] practising a range of techniques of micro-level surveillance as well as enforcement measures; inspecting and examining individuals for scars of vaccination or for evidence of the disease; monitoring and if necessary enforcing movement into quarantined spaces; and encouraging 'healthy' conduct. At a Board of Health meeting, 25 October 1881, it was recommended that some simple directions, based on a 'police code' should be distributed, showing the value of light, air, water, to health.[23] The establishment of an 'Ambulance and Disinfecting Staff' in 1881 under the supervision of the Police (this was the very first action taken by the new Board of Health) further suggests a conflation of policing with health imperatives.[24] Finally, the presence of the Colonial Architect in the Board of Health indicates the contemporary understanding of health and disease as being almost wholly about bodies in architectural and environmental space. Architects, especially public architects understood the design and arrangement of urban space quite literally as an issue of public health, just as doctors in the field of public health or social medicine themselves became 'specialists of space'.[25]

The epidemic was not simply a naturally occurring biological phenomenon, but was deeply constituted as a biopolitical and governmental problem which required not only medical but administrative intervention. Biopolitics is not only a response to a set of conditions, it is itself productive. Biopolitics 'engenders the forms of knowledge that structure these problems and interventions'.[26] 'Epidemic' is always in some senses a bureaucratic and political effect. If, in 1881, epidemic created a public health bureaucracy, it is also the case that a political bureaucracy created the epidemic. That is, it was possible for the colonial government *not* to respond to individual illnesses in terms of a projected pattern of illness in the population; it was possible not to declare 'epidemic'. Rather than being self-evident,'epidemic' was produced and pronounced by government. At least one member of the New South Wales Legislative Assembly challenged this move: '[I am] astonished that the Government should encourage the mania about small-pox; Small-pox [is] as common as the day in every part of the world; [I] dare say that in London you would find 80 or 90 cases within a radius of a mile.'[27] A government medical officer, having just arrived in the colony from London 'where smallpox was a matter of common occurrence', found the government response inexplicable. Later he said, 'it began to dawn upon me what small-pox meant to this unvaccinated community'.[28] 'Epidemic' was not only a fact of individually ill subjects, but also of their interpretation within a pattern of morbidity in the population, knowledge produced in expert realms of medicine and epidemiology, government and bureaucracy.

The epidemic existed in embodied subjects – that is, it was lived – but it also existed as information – that is, it was written. It was a phenomenon of the circulation of disease through individual bodies and through the social body, but it was also about the circulation of information on paper. Official response and practice occurred in two main domains, one environmental and spatial, to be discussed below, the other administrative and epidemiological. Agents of government – the manager of the quarantine station, the doctors there employed, police at various levels, local government – all were required to submit information to public servants of health, who turned this information – itself by no means arbitrary – into statistics, graphs, maps and tables. They wrote and recorded, traced and mapped the disease on paper, created classifications, numbered and counted, projected probabilities. This was certainly an 'uninterrupted work of writing',[29] as each sick or suspected individual was located in actual space, and this knowledge swiftly modified into a form of information recordable in multiple abstract ways: a dot on a map of Sydney, signifying the place and geographic clustering of disease; a line on a graph signifying the cumulative progression of the epidemic over time; a number in a table of figures indicating population morbidity and projected patterns. Each sick or suspected individual represented not one unit of information in the epidemiological system, but was dissected in multiple ways, so that a range of statistics could be forthcoming with particular purposes and effects – age, sex, race, occupation, location, contact points, recent movements, vaccination status, previous medical history.

If a central aspiration of modern government is the health of the population, then statistics about health and disease – epidemiology – is doubly invested as a technology of power. One effect of the technology of statistics, is apparent order and control: 'statistical principles ... tame death, render it controllable and predictable, give it a semblance of order, make it calculable'.[30] Statistics was not a transparent recording of data, but rather constituted a knowledge which was productive, which changed things, which had effects, which was a means by which power circulated through the social body. For example, in Sydney in 1881 the ceaseless collection of information by government was re-circulated as data in the daily press listing new cases and their geographic location, numbers of people removed to, and released from, quarantine, announcement of deaths and so forth. At a multitude of local sites, conduct was modified, recommended hygienic practices pursued, contact and communication policed and increasingly self-policed as a response to the circulation of this information.[31] This 'writing' of the epidemic – its conversion and abstraction into information, statistics, a 'natural history', maps, graphs and figures, was one rationality of government which rendered the epidemic visible and apparently controllable. And it was but one technique for the constitution and management of population.

Lines of hygiene: bodies in quarantined space

If one manifestation of the 'art of government' in Sydney in 1881 was the creation of the epidemic on paper, the 'abstract space' of representation, the other major practice, was the re-ordering of social space. Public health turned on the problematisation of bodies in space, forcing questions of population, overcrowding, the condition and design of buildings into a medical domain. All this came to be articulated in a medical and social scientific language of urban and architectural pathology.[32] Foucault suggested that public health took the medical gaze on the sick individual and enlarged it into a social gaze, the diagnosis of a population in space, thus 'medical space can coincide with social space, or, rather, traverse it and wholly penetrate it'.[33]

Diagnosis and therapy centred as much on architectural spaces as on the embodied subjects in them, or more correctly, on the absolute conflation of bodies-in-space as the medical/governmental problem. The spatial environment of those diagnosed with smallpox was immediately problematised, and was often seen to be the cause of disease itself. While there was some use of the term 'germ' or even 'virus' around the epidemic, colonial medical culture in the 1880s had barely taken on the concept of living microbes as the necessary and specific cause for contagious disease.[34] According to the 1883 Report on Smallpox, the epidemic originated not with a germ or even a person, but with a house: 'No. 223 Lower George-street'. And it finished also at a place – the Sanatory Camp. The immediate official response was to quarantine the house, not the individual. *Where* the disease appeared, in this report, seemed more significant than in *whom* it appeared, although the two were obviously linked. The problematised item was not infected people but 'infected houses'.[35] As we shall see in subsequent chapters, this paralleled maritime regulations in which vessels and ports, not individuals were 'diagnosed' and pronounced unclean:

> The house was promptly placed in quarantine; all unvaccinated inmates who were willing were vaccinated; mosquito net was fixed over the windows ... a guard of special constables, relieved every eight hours, was placed at the front and back entrance, barriers were placed round the premises to keep outsiders from coming into

personal contact with the inmates and premises, and a conspicuous yellow 'caution bill' was put up to warn the public of danger.[36]

The Medical Officer's Return, newly required by the Board of Health, sought the following data, essentially diagnosing first the house, then the house dwellers: 'name of householder; number of families in house; number of floors in house; total number of rooms; approximate size of rooms; general state of house and premises as to repair and cleanliness; general state of furniture'. Then the return moved to the patient's medical history, including vaccination history, date of quarantine, condition, and prognosis.[37] The disease was understood less as spreading from person to person (although this is implicit) as spreading from house to house. The Report of the Royal Commission named five major issues most of which concerned not simply bodies or sick people, but bodies in architectural space – overcrowding, the small size of rooms, insufficient window space, habits of uncleanliness and the impossibility of isolating patients.[38] As Alan Mayne has amply demonstrated, infected houses were inspected and many were later destroyed in programs of slum clearances.[39]

In the city, vaccination accompanied this spatial segregation, drawing attention to the clean/dirty boundaries of cordoned streets and houses. Any contact isolated in their home would, as the *Sydney Morning Herald* recorded it, 'put his or her arm through a hole in the fence at the back of the premises' for the doctor to vaccinate.[40] This involved complex rituals of cleanliness and contamination, which by the early twentieth century became increasingly standard procedures of asepsis in the operating theatre, but in 1881 were the rituals of the public *cordons sanitaires*:

> Take with you two of the boxes provided for the purpose – one marked 'clean', and one without a mark ... Stop a short distance from the patient's house, take the ulster from the box marked 'clean', put it on and visit the patient. Wash your hands after seeing the patient. Upon your return take off the ulster, do it up in the brown paper, and tie the parcel. Put it into the box which has no mark and lock it. The two boxes are to be given to the laundress locked. She is provided with duplicate keys, and will put the parcel direct into boiling water.[41]

Indeed one can trace the rituals of the aseptic operating theatre not to hospital practices (antisepsis, for example) but to the practices of quarantine, mail disinfection and the public *cordons sanitaires* of infectious diseases.[42]

Like a sick body, the city of Sydney was purged in 1881. Contaminants were ejected to a place in which the contamination could be contained (or could appear to be contained), thus brought under some sort of order and control. Initially removed to and detained in the existing Quarantine Station, a new infectious diseases hospital and 'Sanatory Camp' was rapidly proposed and built in another isolated coastal region south of the city. This was partly in response to the 10th Annual Report of the Local Government Board (1880–81), which contained considerable information on recent

Figure 2.1 Rituals of cleanliness and contamination at domestic thresholds
Source: 'Smallpox Outbreak in Sydney', *The Illustrated Sydney News*, 9 July 1881, p. 12.
Courtesy, The State Reference Library, State Library of New South Wales.

English management of infectious diseases, and the establishment of separate hospitals specifically for this purpose.[43] The Quarantine Station and later the Sanatory Camp became sites of official, visible and therefore controllable contamination – the creation of places where dirty matter could be put so that order could be restored.[44] David Armstrong has argued, with Mary Douglas, that rituals of cleanliness and contamination focus attention on boundaries and thresholds.[45] Both the Quarantine Station and the Sanatory Camp were boundary places, situated on the very thresholds of the city. The geography of the Quarantine Station, situated between the ocean and the harbour, was described thus:

> It comprises a peninsula with an area of about 750 acres, completely inaccessible from the water in five-sixths of its extent, and connected with the mainland by a comparatively narrow neck. On the harbour side, to which the ground slopes, Spring Cove affords a safe and convenient anchorage for ships of the largest class in close proximity to the ground occupied by persons subject to the quarantine law, whilst the station is situated at a distance from Manly quite sufficient to ensure the safety of that borough.[46]

Yet the Quarantine Station was not simply a boundary place, creating an unambiguous inside and outside. It was an area – a social space *in* the boundary, internally criss-crossed with lines of hygiene.[47] People removed to the Quarantine Station were not indiscriminately discarded or rejected. Although the quarantine ground was necessarily separated from the rest of the city, virtually surrounded by water, it was not internally undifferentiated. It was, rather, a multiply partitioned space, a site with internal separations, classification and spatial/bodily ordering.

The Quarantine Station was divided in the first instance into 'Healthy' and 'Unhealthy' Ground, sometimes called 'Healthy Ground' and 'Hospital Ground', zones separated by a high fence, with movement and contact closely monitored. The newly cordoned zone of the Sanatory Camp was entirely enclosed 'with a galvanized iron fence' and then 'divided into two portions by a cross fence'.[48] Maps show these areas nominated as Hospital Ground, Healthy Ground and the surrounding bush as Neutral Ground, the latter a sort of buffer zone between the infected site and the supposedly uninfected city.[49] In this artificial and temporary clustering and organising of people, the major social classification was between those infected (unclean) and those contacts, currently healthy but always under suspicion (precariously clean). With the exception of race, in this moment of epidemic, health and ill-health became a primary form of identity, shaped through the significance of spatial placement and classification.

More familiar identities were then constructed and permitted within this new division. The processes of classification and ordering were highly manufactured and blatantly reveal colonial understandings of gender, class, and race and the way these identities mapped onto notions of cleanliness and contamination. The demarcated healthy ground was spatially and socially divided into buildings for white men and women, for white married and single men and women, and for white people of different classes. But even this was not seen to be clear enough. It was recommended by the 1883 Report on the Quarantine Station that there be 'improved means of classification' in which classes of people were even more clearly separated, that buildings accommodating each class of people have a high paling fence separating them 'so that the various classes ... may be separated, and then if strict segregation were maintained, a case of infectious sickness occurring in one group would not necessarily prolong the detention of the rest'.[50] In the Hospital Ground at the Sanatory Camp, there was a three-way distinction, the ordinary ward pavilions, rooms for private paying patients, and two 'ward pavilions for Dark Races'.[51]

The division of 'healthy' and 'unhealthy' subsumed all markers of identity except race. In July 1881, eight uninfected Chinese men were sent to the Quarantine Station. Despite what was otherwise the collection of obsessively precise information about individuals in statistics and tables of returns, these men were consistently de-individualised as the 'eight Chinamen from Druitt Town'. They were not housed in the building for uninfected single men, but in specially constructed tents.[52] Even when not diseased, these Chinese men were not integrated into the 'healthy ground' but were separated and placed under specific police guard. It was one constable's special duty to see that they did not 'mix with the healthy people ... [and] did not escape'.[53] The plans for the Sanatory Camp at Little Bay included a separate 'lazarette' for Chinese patients.[54] By virtue of their race, and irrespective of their diagnosis as 'clean' these Chinese men were already pathologised, already seen to be diseased and to require separation. As I show in Chapter 6, the distinction between unhealthy and healthy was already one which mapped onto racial distinctions: to be 'clean' and 'Chinese' was all but impossible in the dominant racial discourse of colonial Australia.

Vaccination was compulsory for certain people in this epidemic, even though there was never a general compulsory vaccination law in New South Wales. Chinese people were, on paper, to be compulsorily vaccinated. All unvaccinated members of the police force were vaccinated, as were nurses sent to the Quarantine Station and the Sanatory Camp, and prisoners and warders of goals.[55] For (white) people in quarantined houses, consent for vaccination was strongly sought, but it was not compelled. Other people elected to be vaccinated because the procedure allowed them safely (it was hypothesised) into quarantined spaces, and safety if they were already there as suspects. The vaccination of people in the Healthy Ground was issued as a government order. It was enforced by threats of longer detention, and by isolating those who refused from those who assented, in essence a secondary punishment, a further punitive detention.[56] This paralleled the treatment of prisoners' refusal to be vaccinated as a prison offence, subject to further punishment, 'seven days' cells', to take one example.[57]

In the strange space of Quarantine – in particular the ever-precarious 'Healthy Ground' – the process of vaccination ostensibly protected the 'clean' from the dangers over the fence in the 'Hospital Ground', and from the dangers in their midst: people in the healthy ground might well be (what came to be called) 'carriers' within an incubation period. But because of the precariousness of vaccination itself within clean/dirty logics, the procedure did not always offer either a sense of, or an actual security against the disease. In 1881, people in the Healthy Ground strongly resisted the government-ordered procedure but eventually consented. The doctor in charge of the Healthy Ground, Mannington Caffyn, vaccinated nearly forty people from the cowpox pustule of 'a fine, healthy child, with a clean skin and of good parentage' who had just arrived at the Quarantine Station as a contact. To the horror of all, the next day she was covered in smallpox pustules and six days later was dead. 'The sense of fear', the doctor remembered 'was in a great measure due to the fact that we were the outcasts of a community and treated very much as were the lepers of old ... the wildest rumours were afloat ... that wives and husbands were endeavouring to suck the poison from each other's arms, whilst some were openly avowing that suicide were better'. Mannington Caffyn's account is that only one of the vaccinated people got smallpox and none died. Indeed he published the event in *The Lancet* as evidence of the non-transmission of smallpox by vaccine lymph.[58] Others, however, put the death of several of the children in the Healthy Ground down to this contamination by government-ordered vaccination. This case suggests the precariousness of vaccination within the logic of infected and clean bodies and spaces. The contagiousness of the process could itself turn a 'clean' space into an infected one. As discussed in the previous chapter, the preventive process itself could be the means of contamination, and cross the logic of quarantine.

The carceral spaces of public health: government, consent and the liberal subject

The public health strategies of isolation and vaccination are linked historically through this question of compulsion: just what can the state force its citizens to do (and to have done to them) for the protection of others? Public health was a significant domain for the working out of different understandings of the responsibilities and rights of government, of subjects and citizens, and indeed (perhaps especially) of non-subjects and non-citizens. Compulsory vaccination and compulsory isolation raised questions about the legitimacy of government forcing citizens and subjects to submit to either, or both. At issue was the crucial question of the sovereignty of individual embodied subjects, both in terms of the suspension of habeas corpus which public health detention powers were increasingly insisting upon, and in terms of incursions into familial and bodily space which vaccination entailed.

An 1853 Act mandated the vaccination of all infants born in England or Wales. This law hardened in 1867 and 1871 and a conscientious objection clause was finally passed in 1907. Immediately upon the passing of the Imperial Vaccination Act (1853) several of the Australian colonies, which were newly self-governing, followed with their own vaccination Acts. Some were implemented, others dead letter laws.[59] In the same pattern, conscientious objection clauses were inserted into various vaccination and public health laws in the early twentieth century.[60] Partly because smallpox was not endemic on the continent, public response to vaccination was volatile. When there was smallpox around, as in Sydney in 1881, many people consented to vaccination: 68,962 were recorded as vaccinated in 1881, and only 2,188 in 1882, even though about 28,000 children were born in New South Wales in 1882.[61] Opposition to compulsion was intense in the Australian colonies, and in many ways more successful than that in England. But in failing to implement compulsion as strictly as authorities in England and Wales, colonial governments were often swayed by logistical, as much as political, reasons. In many cases in the Australian colonies, it was recognised that the sparseness of the population itself reduced the risk of the disease: 'small-pox does not so easily spread where there is little communication,' wrote the government medical adviser to the New South Wales Colonial Secretary in 1859, advising against compulsion, partly on grounds of cost.[62] Additionally, it was recognised that there simply were not the practitioners available in many isolated areas, or that the distances children and parents had to travel made vaccination, its assessment a week later and revaccination impossible.[63]

Despite these pragmatic considerations, compulsion was also debated in terms of liberal philosophy, as in England. Nadja Durbach has shown how in England there was a strong working-class anti-vaccinationism which politicised compulsion within an existing radical opposition to the New Poor Law and to the Anatomy Act (1831). Compulsory vaccination was understood to be part of a class tyranny and, as she suggests 'by legislating for children, anti-vaccinationists maintained, the state usurped the parental role and disciplined the domestic realm'.[64] In England and the Australian colonies (as elsewhere) pro-vaccinationists countered that because neglecting to vaccinate children put other children at risk, arguments about the need to safeguard voluntariness, personal liberty and the rights of parents to own the bodies of their children, were void. This, indeed, is the core of the significance of communicable disease and its management for liberal governance: it so often requires weighing of risks and benefits for a 'dangerous' individual and for the population-at-risk. For example, the New South Wales Registrar-General discussed with the Colonial Secretary the pros and cons of adopting a version of the Imperial Vaccination Act in that Colony:

> Objections have been, and will be raised to making vaccination compulsory, but its advocates consider it an abuse of the 'Voluntary Principle' to allow a parent not only to risk the life of his own child by neglecting to apply to it what is almost a sure specific against so fatal a disease, but to imperil the health and lives of the community ... indeed to allow a man what is neither more nor less than the freedom to spread disease through the country.[65]

The compulsion at issue was not only about the vaccination process itself, but as Durbach shows, the implications of non-compliance: fines for those who refused vaccination of their infants; the seizing of property of those who refused to or could not pay fines. Ultimately (and not infrequently) compulsion was enforced through imprisonment.[66]

Here again, the medical and penal systems dovetailed, the spaces, institutions and powers of public health, the law and criminal justice crossed-over. Conversely, those already detained on public health grounds in the Quarantine Station in 1881 were ordered by government to be vaccinated at a time and place where compulsory vaccination had never entered into law. As the doctor charged with vaccination recounted: 'a whisper of mutiny ran through the settlement. In obedience to the first instincts of a young democracy a public meeting was summoned and a shoemaker of great verbosity was invited to the chair ... a resolution ... was presented to me ... it entered very deeply into the scientific inaccuracy of the whole theory of vaccination and the gross interference with the liberty of the subject, and ended with a distinct refusal to submit to the operation'. In this case, the 'threat of prolonged detention' itself apparently changed the minds of these objectors, but turned into the nightmare situation where they were all vaccinated from a child who was about to die from smallpox.[67]

In the Quarantine Station, the gaze at work was simultaneously medical and penal. Both the infected people contained in the Quarantine Station *and* those contained there as guards and doctors, were subject to the surveillance and limitation of movement, contact and activity characteristic of the prison, as well as an intensely close medical surveillance. A number of techniques of visualisation rendered the space both therapeutic and carceral. In penal mode, there were permanent guards at entrances and exits, to the Quarantine Station itself and between the 'healthy' and the 'hospital' grounds.[68] Movement and connection between people was thoroughly monitored by police, but also, perhaps more significantly, by people themselves, cautious of the status of those around them. People in the 'healthy ground' were especially observed and self-observing, for at any moment someone declared 'clean' as we have seen, could turn out to be carrying the smallpox. These people were inspected daily by the two doctors, but similarly kept a constantly suspicious gaze on each other. The doctors themselves were simultaneously therapists and inmates, placed without choice in the Quarantine Station because of their early contact with smallpox patients in the city.

What is striking and significant about the 1881 epidemic was the explicit problematisation of compulsory detention, the act of coercively enclosing subjects in the Quarantine Station and later the Sanatory Camp. The episode sharpened the whole question of coercion and consent, which had already been in circulation with respect to vaccination. The nature and legitimacy of isolation powers preoccupied the members of the new Board of Health and the subsequent Royal Commission on the Quarantine Station. The Under-Secretary for Finance and Trade, the Department responsible for the management of the Quarantine Station, stated that 'the Government considered that they had a right under the statute and in the interest of the public health, to remove persons who were infected in the immediate vicinity of infection, from the place of infection, so as to isolate them from the rest of the community'.[69] However in his opinion where people 'objected to go [they] were not pressed to do so'. This contradiction, this official uncertainty about government powers appeared again and again in the evidence and testimonies which made up the Royal Commission after the event of the epidemic. Each official – police, bureaucrats, doctors – denied the use of force, even though it would appear that they were specifically empowered to isolate without consent. The Board of Health itself seemed confused over the question. On the one hand it consistently sought an increase in, and a greater clarification of powers for compulsory removal.[70] On the other, it seemed committed to modes of 'persuasion' and 'encouragement' towards consent. For example, early in 1882 when smallpox contacts were being sent to the Sanatory Camp the Board sought 'consent' thus: 'with a view to encourage the removal of patients and of inmates of infected houses to Little Bay it is desirable that such persons should be informed that only in the event of their consenting to leave their house will the Government make good to them their wages'.[71]

The Royal Commissioners asked constant questions about whether force was or was not used, whether orders from the Inspector-General of Police included the use of force, and about what happened to people who objected to going.[72] The Inspector-General of Police, although initially equivocal, intimated that the use of force was not approved, and finished his questioning with a categorical statement that he 'never authorized the use of force', except, that is, for the removal of the Chinese men from Druitt Town: 'In that case it was after dark, and as the man would have escaped surreptitiously and spread the disease, it was the imperative duty of the police to prevent him [sic] from getting away'.[73] The difference which race made to the questioning or the acceptability of force, was stark. The intricate discussion over the use of force related largely to the whiteness of the majority of those detained. This contrasted dramatically with the relatively smooth passing of leprosy detention powers, to be discussed in Chapter 4.

Many quarantined people gave evidence about their forcible removal and confinement. One doctor inadvertently ended up in quarantine, having treated the first few smallpox cases in their homes and was made to act as medical officer in Quarantine for several months. He told his story thus:

> I went to the Police Station and a reporter from some paper came up to me and said, 'have you heard that you are to go to Quarantine?' ... As soon as my name was mentioned this other person flew across the pavement and ordered me off in the most rough way, and said if I did not take myself off he would give me into the custody of the police. I explained to him that I was quite clean and had no smallpox about me ... He said if I did not prepare at once to go to the Quarantine Station he would have no more nonsense but give me into the custody of the police and have me sent to the station at once.
>
> Do you consider that you were forcibly removed to the Quarantine Station?
>
> Oh yes. I had no option in the matter.[74]

The Superintendent of Police was questioned over the consent of another doctor, also ordered to the Quarantine Station both because he was to act as medical officer and because he was himself a contact. He was asked: 'Did Dr Clune consent to go into the Quarantine?':

> I cannot say that he did; he went; that is all I know; unless the intimation given to him that the Government were determined that he should go to Quarantine, and that he was expected to go, might be construed into a consent on his part; he came to the wharf and he went on board.
>
> But he expressed himself unwilling to go?
>
> I cannot say; he intimated that he regarded it as a grievance and a hardship. He did not say, – 'I don't want to go', in so many words.[75]

One police constable was taken to the Quarantine Station after being diagnosed with smallpox (incorrectly, as it turned out). His evidence to the Royal Commission indicates a confusion between his situation as a constable, required to obey orders, and a patient/inmate with a right to question his diagnosis and to refuse to be sent to the hulk, the

'Faraway', for infected men, or to live in a tent next to the Chinese. Simultaneously guard and prisoner, he was asked if he went to the Quarantine Station under force:

> Yes; I felt that at the time ... A small boat with Mr Carroll [manager of the Quarantine Station] in it pulled towards us, and Carroll asked whether I was a European or a Chinaman ... Then said Carroll, 'I have orders to put you on board the 'Faraway;' you have small-pox.' I said, 'Mr Carroll, I have not, and I don't intend to go on board the "Faraway".' He then went away, and returned after a short time, and I again refused, and said I would not go on board the 'Faraway' where I knew the small-pox really was ... Mr Carroll came with a piece of paper in his hand, and he says, 'Cook, you have small-pox; Dr Alleyne says it, and you must go on board the "Faraway", and I refused again. Then all the constables gathered together, under the jetty ... "Cook, will you go on board the 'Faraway?' " and I said "No" again, and he said, "We have orders to put the handcuffs on you and put you on board"; then I made use of the expression, "By Jesus Christ, I am confident I have no small-pox, and I will sooner die than go on board that 'Faraway'."[76]

When this man was pronounced free of smallpox by another doctor, he was ordered to guard the Chinese men in tents separate from the Europeans. Again, he refused: 'I would have done the duty, but I did not like to go and live among the Chinamen because it was rumoured that they were infected with small-pox ... I did not wish to disobey orders but I wanted to get out of sleeping in the tent ... I said, "I do not refuse to do the duty, but I refuse to go and live among the Chinamen".'[77]

The story told by another man, John Hughes, about the detention of himself, his wife and their six children poignantly gathers together the practices and problems of segregation, vaccination and compulsion which were at work in the epidemic, and illustrates the extent to which detention was both medical and penal. Two constables knocked on Hughes' inner-city door, called 'fire', and when he came down the Health Officer, standing at a distance, diagnosed him with smallpox. Along with his wife and children, he was sent 'by force' and without consent, he told the commissioners, to the Quarantine Station. The family were sent initially to the Healthy Ground, while Hughes was sent to the hulk and placed with 'five Chinamen' in a room with one bed: 'the five Chinamen laid on the floor'.

Later he was placed in a room with a German man who was very sick with smallpox. Hughes also objected to sleeping with him because he was convinced of his own healthiness. 'I won't sleep there any more', 'he told the superintendent of the hulk. Hughes told the Commission that the five Chinamen 'started up and threatened to kill' the superintendent at one point, and then detailed his own acts of defiance. Communicating with his wife on the shore, he repeatedly swam in, was placed in handcuffs on return, but would escape again to see his wife and children. Once he was given permission to go ashore, provided that he not touch his wife or children, and, he recounted, the doctor placed an onion between them at the meeting, 'to keep my breath ... away from them'. By then, his daughter was dying, one of the victims of vaccination-induced smallpox: 'One of my children was vaccinated ... from Mrs Rout's child who had small-pox and got the disease from her', and he escaped again from the hulk. Leg irons were then placed on him and when a policeman presented a pair of 'madman's muffs' he said: 'Are they for me? You will never get them on me, or you and I will have a swim for it'. Hughes' defiance was finally subdued with the use of restraints, and he spent three months on the hulk cooking for the other sick and dying inmates, but never succumbing to the disease himself.[78]

Conclusion

Consent was precisely at issue, both in such micro-encounters on the shores of the harbour and in their retelling in the Royal Commission. It seemed important for the commission officials that their report be able to conclude that no force was used, but that people removed themselves to the Quarantine Station for the sake of the public health, even though this was demonstrably not the case. This difficulty was in part recognised in the act of paying compensation for those detained, in some cases a very considerable amount: for example, £250 to Mrs E. Bonnor, £570 to the family company of the original child On Chong & Co; and an extraordinary £2,515 to Dr Clune. The Hughes family were compensated £20 for detention and £39 for property destroyed.[79]

The 'powers to isolate' granted to the Board of Health were rendered deeply problematical immediately upon their creation. The limitations of that particular public health strategy thus became evident, prompting shifts towards more governmental techniques in which a desire for health and hygiene might be instilled in each and every citizen. It began to be recognised by a whole range of experts and authorities that it was more effective if subjects voluntarily isolated themselves, but this ideal strategy was strongly racialised. Voluntary isolaton occurred, as we shall see, when tuberculosis was newly managed as an infectious disease of the (mainly white) population in the early twentieth century, but not, as I discuss in Chapter 4, in the instance of leprosy. And in the case of suddenly epidemic diseases, like plague in 1900 and influenza in 1919, enforced quarantine continued.[80] The Australian Director-General of Health, Dr J.H.L. Cumpston theorised these questions of law, health and consent constantly. Force may be used to preserve health, he suggested. But, 'as autocratic government no longer exists', that force 'is to be found only in the expressed will of the people':

> Under a democratic system of government such 'will' is ... expressed only in the presence of a universal conviction that the procedure in question is necessary ... So we reach the practical position that in matters of health, law is of small value compared with education to that point of conviction which ensures automatically acquiescent action.[81]

The epidemic of 1881 was one event in colonial Australian history which prompted shifts in the management of population health towards a governmental mode, a mode which was not inevitable or in any way completed, but which was produced in a multitude of local sites in specific response to forceful powers increasingly deemed illiberal. This was one event which shifted public health policy towards the requirement of the 'consent of the governed'.[82] But this consent was sought more diligently of some people, than others in the civic body.

3
Tuberculosis: Governing Healthy Citizens

Smallpox was always considered an 'alien' disease in the Australasian colonies, a disease 'invading' the continent and the population, either through early British contact or through the global movement of Chinese goldseekers and indentured labourers. Smallpox *became* a problem of the white community, but its origins were always comprehended in medical literature as generically Eastern, Asian or Chinese. Tuberculosis, by contrast, was 'the great white plague', a disease originating with and belonging to 'civilized man' as S. Lyle Cummins put it in his study *Empire and Colonial Tuberculosis*.[1] Newly understood as communicable around the turn of the century, tuberculosis endemically disabled populations in industrialised and urbanised countries. As one British expert on sanatorium treatment opened his book on the subject, 'tuberculosis is a disease of communal life ... It is practically unknown amongst wandering and nomadic people'.[2] Epidemiologists in Australia comprehended tuberculosis as a deeply worrying and intrinsic aspect of British or white communities and cultures. Indeed despite tuberculosis being now recognised as a leading cause of Aboriginal mortality in the period,[3] its dominant conceptualisation as a disease of whites almost entirely shaped expert knowledge and management of it. Like smallpox and leprosy, tuberculosis was managed spatially in the early twentieth century. But unlike the coercion exercised against the smallpox contact or sufferer, and as we shall see, of the deeply racialised 'leper', consumptives were enjoined to remove *themselves* to the new institution of isolation, the sanatorium.

My broad concern in this chapter is to analyse the early twentieth-century management of tuberculosis within historical sociology on spatial isolation of the dangerous, and on the formation of the self-governing hygienic subject within institutions. Much historical sociology has been interested in methods of prevention in liberal rule, from older methods of the exclusion of the 'dangerous' to the population-level 'risk' based strategies of prevention which, it is often argued, characterised later twentieth-century mechanisms. In his seminal 1991 article, Robert Castel argued for a qualitative shift in modern strategies of prevention 'from dangerousness to risk'.[4] Castel writes that the question was posed in the nineteenth century: 'how will it be possible to prevent without being forced to confine?' He suggests that governments and experts were finding confinement measures of prevention crude and limited in that 'one cannot confine masses of people', and unresponsive in that the strategy 'can only be carried out on individual patients one by one'. Like many recent sociologists he identifies (and criticises) 'risk' as the major late twentieth-century solution to the problems of prevention-through-isolation.[5]

Both Castel in 'From Dangerousness to Risk' and Foucault in 'About the Concept of the Dangerous Individual' focus their analyses on examples from criminal psychiatry. They argue that the dangerous individual was not necessarily 'diagnosed' as such because of their criminal action, but because of a quality inhering in the individual which suggested that such asocial action *might* happen. What one *is*, not what one *does* became key to the psychiatric and legal response: 'Are there individuals who are intrinsically dangerous? By what signs can they be recognized, and how can one react to their presence?'[6] Unpredictability was the problem to work with, for 'even those who appear calm [or healthy] carry a threat'.[7] Thus, in order to *prevent* the threatened pathological act the dangerous individual needed to be diagnosed as such by increasingly specialised medical/psychiatric experts, and confined in a prison, or an asylum. Isolation was a preventive measure, rather more than a punishment measure: 'to confine signified to neutralize, if possible in advance, an individual deemed dangerous'.[8] For white people, some characteristic beyond their race constituted them as dangerous. For Asian or Indigenous people, as we shall see in the next chapter, their race alone often constituted them as dangerous.

While there is very little medical or public health history which looks at prevention in terms of this sociological literature,[9] it is clear that the dangerous individual was also the infected but symptomless person, the new category of the 'carrier'. The danger of the symptomless contact drove much policy on smallpox management, as I have discussed. But from about 1890, the newly termed 'carrier' acquired great significance and posed major new problems for government.[10] In part this was driven by new possibilities of bacteriologically diagnosing a carrier, rather than waiting for the possibility of diagnosing clinically, the symptomatic. 'The germ carrier and the law', as one article summarised the problem, posed a conundrum for liberal governance: 'can a person carrying a germ but who is not sick, be quarantined against his will?'[11] In many instances, the answer was yes. The isolation of 'Typhoid Mary' in New York, for example, was an extraordinary episode produced by the new problem of the dangerous carrier.[12] The threatened pathological act in that case was not violence, but the spread of contagion. In the way that the criminal individual *appeared* calm, the carrier appeared healthy, but both were nonetheless dangerous. As I show in this chapter, 'danger' and 'dangerousness' was not infrequently the precise language used by experts and governments with respect to tuberculosis, as the safety and health of the population became increasingly an object of intervention.

The common legislative and management innovation of the late nineteenth century was the technique of notification, which had a twin – isolation. Such powers were classically implemented in the management of venereal diseases under British and colonial Contagious Diseases Acts.[13] In Britain, notification incrementally increased from the middle of the nineteenth century, with some 'setback' after the public reaction to the CD Acts, but then mushroomed after the 1889 Infectious Diseases (Notification) Act and the Act's significant intensification in 1899.[14] Measures for compulsory notification were thus not in themselves new, but they expanded prolifically in this period, and were accompanied by all kinds of novel powers for the compulsory examination and detention of the infected. To take the instance of the British colony and later Australian state of Victoria, leprosy was first notifiable in 1893, tuberculosis in 1909, diphtheria in 1916, puerperal fever in 1917, polio and malaria, in 1920. In New South Wales, diphtheria was made compulsorily notifiable in 1898, bubonic plague in 1900, tuberculosis in 1904, infantile paralysis in 1912, enteric fever in 1921.[15] Although the reasons for the introduction of notification in specific years in specific locales are multiple (particular governments, particular epidemics, particular diagnostic possibilities), notification proliferated without precedent in the turn of the century period. So did places of isolation. There were precedents for 'isolation hospitals': early modern syphilitic hospitals; separate maternity wards for women with puerperal fever in the nineteenth century; and Lock Hospitals for prostitutes with venereal diseases which in many ways stood as the modern template for the exclusion of the undesirable. Yet, alongside notification, both the measures to enforce isolation on health grounds, and the number and specialised kind of places of isolation were newly expanding. Bacteriology offered both better diagnostic tools and new rationales for the location and confinement of the (medically) dangerous, and progressivist and liberal-welfare states created new imperatives for the pursuit of bureaucratic information on individuals and populations. Spatial isolation was far from the *only* public health response, but what is significant is the expansion and refinement of spaces of isolation which clearly characterised this period. This is the period of the expansion of the leper colony, institutions for the feeble-minded, the epileptic colony, and, as I discuss here, the sanatorium for consumptives.[16]

At the same time, though, the period was also characterised by the evolution of hygienic self-governance. The case of tuberculosis management in sanatoria stands as a perfect example of the simultaneity of these measures, both of isolation-of-the-dangerous and modes of hygienic self governance more characteristic of 'the new public health'.[17] Sanatoria are also significant because consumptives often entered the institutions voluntarily, albeit for a range of reasons and some with little choice due to their severely impoverished state. Once 'inside' the aim of open-air treatment in sanatoria was to reform consumptives into responsible self-governing, non-infective ('safe') and hygienic citizens. The sanatorium was one institution where the 'soul of the citizen' came to be intensively governed.[18] It was a deeply heterotopic space, embodying multiple traditions and meanings of segregation and institutionalisation – penal, charitable, preventive, educative, restorative and therapeutic.

Here, then, I analyse early twentieth century tuberculosis management within literature on isolation, public health spaces, and management of the dangerous. In the first sections I show how the shift towards understanding tuberculosis as communicable, and therefore as an issue of *public* health was accompanied by the language of 'dangerousness'. This drove new strategies of institutional and isolated therapy and prevention. In the third section, I discuss the instruction and training which went on within these isolated spaces, as not only (white) bodies, but souls became the project of the experts, and, significantly, of the consumptives themselves. And finally I discuss healthy citizenship and civic responsibility to be 'safely' released back into the community, as part of this new cultivation of the hygienic self. Throughout, I explore isolation in the turn-of-the-century sanatorium as situated *between* coerced confinement, and 'voluntary' confinement, which itself signalled new modes of governance-through-freedom, increasingly available to white subjects.

Tuberculosis prevention and treatment: toward 'public' health

Throughout the nineteenth century treatment practices for tuberculosis, or consumption as it was then called, relied primarily on an environmental aetiological and therapeutic model, one in which climate was especially important. Consumption was considered responsive to qualities of the air – purity, temperature, altitude, and/or proximity to the ocean. Direct medications were certainly prescribed for the treatment of consumption – variously strychnine, sal ammoniac, mercury, antimony, and cod liver oil – as were a range of standard Victorian therapeutic practices, such as the application of counter-irritants and various kinds of bathing.[19] Nonetheless, these were usually prescribed as adjuncts to some version of environmental and climatic remedy. Australia, like many 'new world' places, was intermittently presented as ideal for the recovering British consumptive. Isaac Brown, in *Australia for the Consumptive Invalid* (1865), wrote of Tasmania: 'Sea-breezes on all sides bring *pure air*, disinfecting as it were the whole island'.[20] The sea-voyage between England and Australia was itself often understood to be curative, although as with almost every facet of therapeutics and medicine in the mid to late nineteenth century, each of these claims was disputed.[21]

'Open-air' treatment in sanatoria, which became common in Europe, Britain, North America and Australia at the turn of the century, was no radical departure, but rather was derivative of the dominant climatic understanding of consumption. It was one manifestation of this established discourse of health and ill-health, albeit especially organised, systematic and institutionalised – that is, modern. Modelled on German private institutions for open-air treatment, the first British sanatorium was established in Edinburgh in 1889, and as Linda Bryder has demonstrated, there was a phenomenal proliferation of institutions for the prevention and treatment of tuberculosis (not all of them sanatoria) over the turn of the century. In 1886 she notes 19 specialist hospitals in England and Wales and in 1920, there were 176 institutions.[22] In the United States, the sanatorium treatment was popularised from about 1884 and Michael Teller notes a similar mushrooming of institutions for various kinds of open-air treatment.[23] In Australia, there were some charitable institutions for consumptives from the 1880s, but open-air treatment in private and public sanatoria was institutionalised from the beginning of the twentieth century. The focus on removing patients to therapeutic locations and climates, the long sea voyage as remedy, as well as the earliest private health resorts/sanatoria in the mountains of Europe, indicate the middle and upper-class focus of preventive and curative strategies. None of these practices were initially 'public' preventive strategies in any way, both in that they were not directed to working-class people, and in that they were aimed at individuals not populations. But there was a discernable shift around the turn of the century, noted by many historians, in which focus moved from the individual and often aestheticised consumptive towards an imagining of the disease as one of urbanisation and industrialisation, a disease of the working class, of the 'public'. For example Katherine Ott notes with respect to the United States, that the comprehension of tuberculosis changed from being an 'allure of delicate consumptions' toward association with 'poverty and the dangerous classes'.[24] In the United States this shift was strongly racialised, with African-Americans targeted and, it has been argued, segregated in urban spaces.[25] In Australia, the problem was strongly, almost exclusively, comprehended as a disease of whites.

Accordingly, sanatoria which had emerged as private institutions, soon became public institutions as the management of tuberculosis became entwined with public health mechanisms and with early welfare states. Indeed tuberculosis became key in the development of government welfare responsibilities not only in the management of emergency epidemics, but also in the control of diseases which threatened the population and the economy in chronic ways. Not only in Australia, but importantly in Britain and Germany, tuberculosis was a crucial problem through which newly welfarist governments began to assume responsibility for both treatment and prevention, and through which sickness benefits, invalid pensions and health insurance measures began to be implemented.[26]

If smallpox had long been understood as the classic contagious disease, tuberculosis was only gradually comprehended thus after Koch's isolation of the bacillus in 1882.[27] Historians differ over the significance of the isolation of the bacillus for public health strategies to prevent tuberculosis. Some argue that is was key, that it launched a 'new era of prevention'.[28] Others see it as important bacteriologically, but more or less irrelevant to the dominant movement towards open-air prevention and therapy.[29] The principles of open-air sanatorium treatment existed well before the general acceptance of germ theories or the revision of medical and preventive cultures towards a bacteriological model. Therapy based on sunshine, fresh air, climate, diet and exercise could be reworked to incorporate bacteriology, but certainly did not develop from it, or rely on it to make sense. But my concern here is not to explore how bacteriological developments in aetiology after Koch's isolation of the bacillus displaced or merged with pre-existing clinical management or prevention. Rather it is to explore the significance of the new comprehension of tuberculosis as *contagious* or really infectious (as it was indirectly transmitted) for emerging *public* health, its institutions, responsibilities and instrumentalities.

A clear statement of the significance of rethinking a disease as infectious was made at a public meeting in the Sydney Town Hall in 1901. There, tuberculosis control was presented as a new duty of the State *because* it was an infectious disease threatening the community. Citizens should therefore expect protection from danger:

> In order to prevent a man or a woman who had been poisoned infecting with that poison somebody else, it was clearly the duty on the part of the State to make provision for the exclusive treatment of those suffering persons who were at present a source of danger to themselves and to all around them ... Since consumption is an infectious disease, it is the duty of the State to adopt reasonable and effective measures for the prevention of infection.[30]

And again in 1906 a deputation from a Municipal Council's meeting was received by the Chief Secretary. They put forward arguments first for the compulsory notification of consumption, and secondarily, for an institution into which those with consumption could be put, not for treatment, but as a quarantining measure to protect the State's citizens: 'provision of some institutions where they might be segregated from the main body of the people of the State. By those means only could the health of the people of the State be protected.'[31] They were clearly drawing on emergency quarantining measures of smallpox or plague isolation as the model for the management of tuberculosis as an infectious disease.[32] But epidemiologically the diseases were very different. Tuberculosis was not an acute epidemic diseases, but a chronic, endemic one. It was comprehended as a 'core' disease of middle-age, of white domestic spaces, and of the workplace. Repeatedly in medical, government and popular literature, it was noted that tuberculosis was a disease which struck men in paid employment and women who were working, bearing children and raising them. Mortality in New South

Wales was highest for both rural and urban white men in the 30–39 age group and for urban and rural white women in the 20–29 age group.[33] It killed people in the prime of working life, and 'during the marriageable ages',[34] deeply affecting the core social structures of 'family' and 'labour'. Entire family economies were disabled through the chronic and endemic nature of the disease.[35]

Despite the differences with the diseases against which isolation had traditionally been used, the New South Wales government was persuaded of its duties to prevent the spread of infection and built its own institution for the isolation of consumptives at Waterfall, south of the city, in the same general area as the 1881 Smallpox Sanatorium, now the Infectious Diseases Hospital.[36] It was specifically a response to the new conceptualisation of the consumptive as dangerous because contagious. On this model, the sanatorium was more a quarantine station or an isolation hospital than anything else: the crude segregative move of a *cordon sanitaire* protecting the rest of the state's citizens. As a *public* institution it was mainly for the indigent, the working-class, or middle-class people impoverished because of the disease. In some ways, the consumptive sanatorium drew from the idea of the government asylum, providing shelter and care for 'incurables', different to, but in the tradition of, the English workhouse.[37] But this meaning of segregation quickly mapped onto other rationales for institutionalisation and isolation. Over time both the 'quarantine' and the 'asylum' versions of the sanatorium shaded almost imperceptibly into the sanatorium-proper where long-term, expensive and highly individualised treatment regimes were undertaken.

Isolation and the dangerous consumptive

'Danger' and 'dangerousness' was the vocabulary of Australian health and hygiene experts when they turned to consider tuberculosis as a problem of public health: 'Every consumptive [is] a source of danger', announced a 1911 Report on Consumption.[38] In 1909 one doctor asked what was to be done with the consumptive who 'cannot manage himself, and is a perpetual or intermittent source of danger to his neighbours?'[39] And the editor of Sydney's *Daily Telegraph* warned that consumptives were 'at present a source of danger to themselves and to all around them'.[40] This language of dangerousness both recommended and justified new institutional isolation among the preventive responses. When the state health ministers met over the issue of consumption in 1911 they recommended five measures to be implemented uniformly, based on the creation of new legal powers: compulsory notification; legal powers to regulate the home management of consumptives; legal powers to 'remove dangerous or infective consumptives into segregation'; powers to detain them in segregation; legal power to medically examine contacts of consumptives.[41] In the management of tuberculosis, then, one can see not only the retention of sovereign powers of removal and detention which had been used in epidemics of various kinds, notably the smallpox epidemics, but their refinement and extension to other kinds of 'dangerousness'.

But there were differences between 'dangers'. Unlike the case of smallpox, leprosy or plague, where having the disease or being a contact was enough for the powers of isolation to be enforced, diagnosis with tuberculosis alone did not warrant similar immediate action. Rather, isolation in the case of tuberculosis was reserved for people who were proving ungovernable in some other respect as well, usually working-class people without identifiable homes which could be inspected and sanitised by a visiting nurse or a sanitary inspector. Their indigence meant invisibility in a public health system which relied on spatial tracking, their lack of place was understood as dangerous 'roaming', spreading the disease in unknown ways as they moved uncontrolled and unmonitored through the city. This was a longstanding classed understanding of the management of those who could not responsibly govern themselves.[42] The ungovernable consumptive who could not manage himself, needed isolating. His whole being was understood as infective and dangerous, and therefore he needed to be managed totally. The Melbourne psychiatrist and eugenicist J.W. Springthorpe wrote in 1912:

> The danger, the greatest danger of all, is from careless, generally advanced patients, walking about at large, using ordinary handkerchiefs, and spitting here, there, and everywhere ... Such patients should be taught the danger, that it affects themselves, also, and how to cease being a danger ... in many cases, especially among the poor and ignorant, the sufferers must be aggregated into suitable homes or institutions which need not be dangers to others.[43]

The new contagiousness of tuberculosis made these ungovernable people dangerous. Once public dangers, the government had a further rationale to regulate their conduct and to secure the safety of others: the medically dangerous could be criminalised in the new discourse of the carrier. 'Such actions should be regarded as a grave crime against the community', wrote Springthorpe. And for Commonwealth Director-General of Health, J.H.L. Cumpston, 'Spitting at all times is a disgusting practice, but when it is a method of dissemination of tuberculosis, then it becomes criminal'.[44] There was a sense, then, in which the government sanatorium was in part a carceral space, where penal codes as well as health codes could remove people, where those posing dangers could be prevented from doing so, or even further along the carceral continuum, were *punished* for doing so.

There was considerable discussion about the nature of the dangerousness of the consumptive. That is, did the dangerousness inhere in the person him or herself, as an integral aspect or element of the person, a 'quality of the subject', something like the criminal personality.[45] Or was the dangerousness arbitrarily attached, a substance if you like which could be detached. On the one hand the new concern about microbes located danger 'everywhere'.[46] On the other hand, bacteriology offered ways in which to de-pathologise, de-psychologise and in many ways de-mythologise, the consumptive subject, and to locate the dangerousness not as an inherent quality, but, simply as identifiable microbes in the body. The 1911 Report argued thus:

> In what does this dangerousness consist? Only in the expectoration they give off. Their dangerousness is proportioned to the amount and virulence of their expectoration ... the actual danger ... depends entirely on the care they take of their expectoration. If they collect and destroy it, others are in no danger ... but if they spit about carelessly, and soil their clothes, floors and handkerchiefs ... the danger to others is great.[47]

Minimising danger, then, was sometimes about isolating the unmanageable in institutions, but it was also about teaching certain habits, certain safe and responsible modes of conduct. Consumption, the ministers reported 'can be easily managed with safety'.[48]

> [S]egregation of the sick from the healthy should be our aim. But it would be unnecessary, useless, impracticable and improper, to advise segregation as a routine measure to be inexorably carried out in every case. All consumptives are not dangerous; few consumptives are dangerous throughout their illness; and even those consumptives who are most dangerous can surely live among the healthy with safety to them, by punctual observance of simple and easy precautions.[49]

Some resisted the shift towards isolation-as-prevention for tuberculosis, arguing within a liberal paradigm about freedom of movement, and the right of white citizens not to have their intimate lives regulated by the state, in the tradition of anti-vaccinationism, Contagious Diseases Acts agitation, and opposition to emergency quarantine discussed in the last chapter. Notably, such arguments opposed isolation as the undesirable ramification of a 'contagionist' doctrine. One commentator wrote: 'According to their [the contagionists'] recommendations, such sufferers are to be deprived of the rights of citizenship. Their comings in and goings out are to be strictly prescribed by laws. They are to eat and sleep in solitude, and as far as possible they are to have no communication with their fellow-creatures'.[50] Such powers were rejected as 'medieval' and commentators not infrequently resorted to the story of European leprosy isolation: 'In short, many of the extremists who advocate this doctrine of contagion, would isolate the unfortunate victim of phthisis as effectually, and with as little compunction, as the leper of the Middle Ages'.[51] Indeed leprosy and tuberculosis were not infrequently connected, not least by Robert Koch himself, as individuals, states, experts, and sufferers sought ways of understanding the best means of prevention. Koch argued that consumptives, like lepers, should be isolated. But in Australia the coercion involved in the case of the sanatorium for consumptives was ambiguous, far more so than for the smallpox victims of 1881, and in fact bore little resemblance to practices from 'the Middle Ages'. Indeed as we see in the following chapter, there was no need to resort to 'the Middle Ages' to illustrate coerced and lifelong exclusion as both New South Wales and Queensland had passed Leprosy Acts in the early 1890s, newly allowing for (that is, requiring) such detention.

For working-class white people and people unable to work at all, the government and charitable sanatoria were kinds of government asylums, in some cases providing the only option for care and residence. For middle-class people the institutions could be imagined as a health resort in which the discipline was part of the therapy voluntarily entered into. Most consumptives could in fact leave the sanatorium if they wished: unlike people with leprosy or Aboriginal people on protectorates and reserves, their freedom of movement was not severely curtailed. Thus while the language of dangerousness was accompanied by a proliferation of legal and regulatory powers, what differentiates the management of tuberculosis from the management of epidemic diseases like smallpox or plague was the voluntary nature of the segregation practices. For the most part, isolation in this case occurred with 'the consent of the governed'.[52]

The sanatorium: cultivation of healthy selves

The sanatorium was a strange, hybrid place with a peculiar genealogy. If one lineage was represented by the European health resort and late nineteenth-century middle-class 'rest-cures' and other therapies for neurasthenia,[53] the sanatorium also derived from a lineage of working-class reform institutions. Sanatoria certainly emerged in both private (middle-class) and public/charitable (working-class) versions. Yet I am interested to explore the unlikely way in which the two traditions merged, both in the sense that such lengthy and individualised (expensive) treatment was taken up as *public* health at all, and in the way that the private sanatoria were so deeply disciplined in a reformatory model. In her study of tuberculosis in twentieth-century Britain, Linda Bryder argues for the proliferation of sanatoria as part of a culture of institutions and institutionalisation.[54] It is worth pursuing her insight by asking further questions about what institutionalisation meant, as a practice of isolation of the undesirable and dangerous, as prevention-through-confinement, and as a practice of reform and correction. Sanatoria were interesting places indeed: 'total' yet voluntary, places for the isolation of the dangerous yet also for their instruction and training, highly policed but aiming to produce self-policing subjects, places of both therapy and prevention. In short, while sanatoria are recognisable as classic disciplinary institutions, their genealogy is ambiguous and multiple: part hospital, part prison, part school, and in their various classed public and private versions, part asylum, or part health resort.

What constituted 'open-air' treatment in a sanatorium? Simply, patients were to spend as much time outdoors as possible. That rural 'outdoor' place was carefully located by experts, according to ever-changing opinion on the benefits of mountain-air, sea-air, temperature and prevailing winds. Activities usually undertaken inside – rest, sleep, schooling, eating and so forth – were to be transferred outside wherever possible; hence the vogue for 'open-air schools', for example. Most emphasis was placed on breathing 'outside' pure air through the night.[55] Thus consumptives were urged (and in the case of sanatorium patients were required) to sleep in tents, or on verandahs or in the peculiar innovation of the period, the consumptives' chalet. The chalets were designed to accommodate a single patient, in which air constantly surrounded the tuberculous person, day and night, summer and winter, in virtually any weather, including rain.[56] They were open on all sides, and as one inhabitant put it: 'Being open on four sides and constantly bathed in sun and air, the chalets are pretty well sterilized without any formalin'.[57] The verandah remained a staple of hospital architecture through the first half of the twentieth century largely because of the perceived value of pure outside air. Conversely, the peculiarities of Australian domestic architecture, in which verandahs were common, were consistently praised as especially health-promoting.

Open-air treatment usually meant a disciplined program of rest and exercise. As in many kinds of institutions, from the military to the religious to the therapeutic, eating and rest marked the temporal divisions of the day. Beginning with total bed-rest and under very close and strict medical and nursing direction, the consumptive would incrementally increase their daily regime of exercise and exertion. Initially undertaking short walks at a very slow pace, this might increase over some months to walks of several miles with substantial inclines. 'Before meals we lie on lounges silent for one hour ... After meals we lie again on lounges, like gorged boa-constrictors, for half and hour, and then, if ordered walk out at a snail's pace. When I first saw the patients creeping about, I pitied the poor feeble creatures, but found that it was regulation pace, and I was frequently pulled up for my jaunty tread'.[58] As I discuss below, the public, charitable and working-class sanatoria, sometimes called 'industrial sanatoria', reformulated this as 'graduated labour' and required patients to undertake work as therapy, for example gardening or scrubbing floors.

Diet was carefully regulated, with many institutions forcing large meals on consumptives, seeking significant weight-gains as evidence of recovery. As a precursor to late twentieth century cognitive behavioural therapy for anorexics, all the consumptive's activities were monitored, all rewards were granted or withdrawn according to their weight as well as their normal or abnormal temperature. One consumptive wrote in 1907: 'There is a regular rule about temperature. If one is 98.8 early, must stop in bed; and if 100.2 at 5 p.m., must retire ... I wanted to stay in bed for breakfast, and couldn't (without a regulation temperature) ... Four times a day we take our temperatures, and we do the same thing every day'.[59] Consumptives were to learn all kinds of new practices, new ways of being, in a re-education programme usually recommended at six months.

Sanatoria were 'total institutions' in that every aspect of the patient's life was regulated and monitored.[60] Both temporally and spatially the consumptive was entirely directed, and was to follow a precise series of verbal and written rules, about bodily conduct and interaction, as well as the imperative to live out of doors as much as possible. For working-class consumptives, rule-following in institutions was part of an established class dynamic, reworked in the sanatorium as therapeutic. For example the state-run Greenvale Sanatorium in Victoria presented its rules in a manner similar to any public institution of the period: 'Any patient found communicating with another patient in his ward or tent by speech, signs, or writing, or found moving about or otherwise disturbing the quiet of the tent during the silent rest period is liable to immediate dismissal'.[61] Yet for a consumptive, silence and compliance with the regime was not *just* about maintaining institutional order, it was a specific aspect of the treatment. For middle-class consumptives who might have the option to undertake a version of open-air treatment at home, it was the imposed discipline of the institution which recommended it, which aided one in self-discipline. Thus a 1908 medical congress was told: 'In a sanatorium, the close relation of patient and physician is especially conducive to recovery, and where a good result so often depends on strictly following out orders in apparently trivial details. It is the regularity and precision that counts for so much, and are only to be gained as a rule in a sanatorium.'[62]

Within the sanatorium system, the consumptive was imagined as a *tabula rasa* project for reform, from which the new strong and responsible embodied self could be built. The discipline and excessive order of the sanatoria not only represented conventional institutional culture, but was also a way of re-making the consumptive from first principles, as it were. In this system, total bed-rest and silence constituted a symbolically blank and neutral starting point. The will of the patient was to be neutralised and replaced by the will of the doctor and the institution. Ultimately, and almost necessarily through struggle, the patient would will herself into new modes of conduct, and a new hygienic subjectivity. The consumptive's freedom was removed (or voluntarily forfeited) – 'we do the same thing every day and have no will of our own'[63] – and it was to be re-gained or earned back in an incremental process of instruction and acceptance of a new way of life, in developing 'the will to persevere and beat the microbes'.[64]

At the most mundane but certainly not unimportant level, this meant being trained in precise new bodily habits. The minutiae of bodily conduct – coughing, chewing, kissing, washing, sleeping – was all considered reformable in the interests of the patient, and governable in the interests of the public. The reform of consumptive conduct was centred on the mouth and spitting. But if 'Don't Spit' was the simple order on public health posters of the period, instruction for consumptives was far more detailed:

> Coughing is advantageous only when it enables a patient to bring up phlegm. Other coughing is harmful and exhausting ... When coughing or sneezing hold a rag lightly in front of the mouth or nose in order to catch the cough spray ... Do not stand close to and face to face with the person to whom your are speaking [sic]. Never blow on to hot milk or on any other food substance in order to cool it before taking it ... Do not swallow any spit which comes up from your lungs or which comes from the back of your throat.[65]

And while spitting was the primary focus for the reform of habits, not only for consumptives but for the general public, it was certainly not the only bodily habit under scrutiny. For example, in 1912 the medical officer at the Greenvale Sanatorium in Victoria instructed thus: 'Do not kiss or allow yourself to be kissed on the mouth'. And as for eating and resting:

> Always wash your lips and hands, and rinse out the mouth, before eating ... food must be eaten slowly. It is most important that *all* food, even such articles as porridge or bread-and-milk, should be thoroughly chewed, in order that the saliva in the mouth may be mixed with the food, and the digestion thus, as far as practicable, assisted. Half-an-hour at the very least must be spent over each meal ... Rest, especially before meals, should be as complete as possible, and is best obtained by lying at full length on the bed or couch, and refraining from talking.[66]

There was certainly a notion that eating in the right way, resting and being surrounded by pure air had concrete effects upon the body. But there was an important sense in which the discipline itself, succumbing to rigidly controlled institutional life, and subjecting oneself to certain hardships of 'nature', produced therapeutic 'character' and 'strength'. Discipline and struggle produced stronger bodies with a greater capacity to resist and contain the effects of the disease. Like a range of quasi-military, character-building, 'outdoor' cultural institutions such as the Boy Scouts both the body and soul were tested and invigorated.[67] European 'open-air schools' for children with tuberculosis capture many elements of this culture perfectly. Children were required to undertake sporting drills outside summer and winter, and institutions located in the snow were understood to be especially effective. Adults and children alike were asked to wear minimal clothing in the outdoors to enhance the therapeutic effect. In that tuberculosis was the opposite to a 'tropical' disease like leprosy, this 'cold' therapy can be understood as the therapeutic counterpart to the heat and humidity which was understood to cause disease in whites in the tropics in the first place.

This ascetic and Protestant understanding of struggle and discipline as beneficial for the body and soul was conceptualised as much psychologically as physiologically. The significant term 'resistance' crossed-over between the two domains. Open-air treatment for tuberculosis was understood to build up a physiological resistance to the disease, which at most would eradicate it from an individual's body, and at least would keep it contained so that it incapacitated that person minimally. This general idea of resistance had any number of technical explanations, as the notion of immunity was itself being invented and revised in the period. The 'resisting power' and 'susceptibility' of individuals was much discussed. This was the 'soil' of the 'seed and soil' metaphor for comprehending differential infection:

> The constitutional conditions which render man susceptible are any of those diseases or habits which lower the general tone and diminish resisting power ... to prevent the spread of tuberculosis we must endeavour to secure three things: (1) An atmosphere free from the bacilli or tubercle; (2) an invigoration of the body which will enable it to resist infection; and (3) a healthy environment.[68]

In much of this usage, 'resisting power' was a non-specific notion implying a general constitutional healthiness, but one which was fought for, as it were, attained and earned ascetically through some kind of difficulty. Even when 'resisting power' was located specifically, for example as action in the blood, the imperative to 'keep the blood pure' translated into an imperative vigilance over one's environment, one's program of rest and exercise, one's access to pure air day and night.[69] Often resistance implied the capacity to minimise the disease once infected, as well as the capacity to resist the disease altogether. Conversely these 'pure' and disciplined modes of conduct were understood to somehow translate into a process or substance which attacked germs in the body: 'Attention is now chiefly directed to the search for means of stimulating the special resisting powers of the patient to tuberculosis, and it is commonly assumed that these defensive reactions are directly or indirectly germicidal.'[70]

'Resisting power' was argued by some to be the effect of various tuberculin preparations, in which case vigilant attention to open-air treatment was less important than regular injections. For others the idea of 'auto-inoculation' explained the importance of the constitutional program of rest, exercise, pure air, and diet. But for most experts and patients, enhancing 'resisting power' meant working on character and attitude as much as bodies. One successfully treated advocate of the system wrote: 'A stiff-upper lip is needed, a certain amount of brains are needed, character is needed; a little money is needed; also a mind and imagination kept pure from contaminating thoughts and desires'.[71] Open-air treatment was a 'hardening' treatment.[72] And still another consumptive wrote: 'No pampering. Everything to harden one and make one independent'.[73] The treatment regime, then, was not only about forming certain bodily habits, it was about forming a resisting character, cultivating a strong self out of struggle. The more that psychological discourse became available, the more it integrated into the purpose of a sanatorium: 'A sanatorium is not an institution, it is an atmosphere'.[74] By 1924, one Australian expert Gordon Hislop argued strongly for far greater emphasis on 'the psychological side of the tuberculous patient's character'. At the very least, he suggested, doctors and public health administrators should note the palpable difference in the 'atmosphere' between the graduated labour programs and 'the old system of monotonous rest'. He wrote that 'there is a psychological advantage in that the improvement is made visible by tangible steps'.[75] A sanatorium regime, then, was conceptualised as much as a 'mental' treatment – mental hygiene – as a physiological treatment. As we shall see in the next chapter, the difference between the institutional management of the consumptive and the leper could not be greater.

The significance of institutional discipline for the middle-class consumptive in a private sanatorium is clearly articulated in a fascinating set of letters from a Victorian institution, published in Melbourne in 1907. The anonymous author of *Letters from a Sanatorium* wrote: 'I am in a Reformatory ... hedged with bye-laws, where the days are cut into lengths for rest and exercise, with intervals for temperature taking and meals. The Doctor's word is law'.[76] At another moment she described it more as a convent: 'here in these conventual precincts I am restrained, guarded, protected, preserved'.[77] The shift between imagining herself in some kind of corrective institution on the one hand and a religious community on the other is not accidental. The private sanatorium functioned well within a culture of total order, obedience and punishment in the form of withdrawal of reward, characteristic of both the prison and the ascetic religious community. For her, the sanatorium also had the ambiguity of the asylum, both place of protection-because-vulnerable and place of isolation-because-dangerous. She wrote with a certain irony and humour about life in the sanatorium, and with a fascinating insight into her own 'institutionalisation', as it came to be called some time in the anti-psychiatry moment. 'My world seems bounded by Sanatorium hills,' she wrote. 'I tremble at the thought of going forth into the wide reckless world without the Doctor to direct my steps from the time I open my eyes at dawn until I shut them again at night'.[78] She was very much aware of the deliberate process of re-making herself, of allowing herself to be re-made, which was precisely the aim of the treatment. She described another patient who had become so totally dependent that he could not leave the institution, even when pronounced well: 'I am not as frightened as Mr Bunny is. He has been here five months, and is perfectly well, but cannot make up his mind to leave ... Mr. Bunny hugs his chains'.[79]

Mr Bunny was a failure of the sanatorium system because even though he was technically and physically cured, he had not become a self-policing 'independent' consumptive. The anonymous letter-writer, on the other hand, was a success because she left the institution as a fully self-monitoring convert to 'open-air' life, having internalised the sanatorium's instruction in a new mode of conduct, having developed a new sense of self, even as she had all kinds of insight into this very process. She understood she was not only the 'project' of the doctor and nurses who ran the sanatorium, she was also her own project. She had to cultivate a new consumptive self and with that she could return to her family as safe and responsible mother, and to the community as safe and responsible citizen.

As Rose has put it: 'The citizens of a liberal democracy are to regulate themselves ... to be educated and solicited into a kind of alliance between personal objectives and ambitions and institutionally or socially prized goals or activities'.[80] Of course, as in the case of Mr Bunny, this dream of social government did not always, even rarely, eventuated. But what is important, and what the case of the sanatorium illustrates, is the extent to which experts and authorities actively sought this self-regulation as a prime aim and method of the pursuit of public health, often specifically in contrast to the compulsions of the past. The desire for health was encouraged in individuals. Isolating oneself as dangerous to others, and dedicating oneself to the discipline of open air treatment was not necessarily forced but voluntarily entered into as a responsibility of citizenship. One public health administrator urged that 'the spirit of the age is more that of persuasion'. The aim of public health was to create a situation where 'the impulse of the individual to do the right thing shall arise in his own mind, the "voluntary impulse" of a free man'. This was pronounced not only the age of 'auto-inoculation', but also 'The Age of Auto-Suggestion'.[81]

Hygienic citizenship

Civic responsibility was one of the major attributes which sanatorium training aimed to instill as a core part of the consumptive self. Since neither civic status nor responsibility were usually attributed to non-white people in this period, the entire regime of treatment was premised on the whiteness of consumptives. When the Australian states met in 1911 on the question of tuberculosis they concluded that preventive measures should be based on 'the education of the consumptive, and the awakening of a sense of responsibility in himself'.[82] This need for responsibility was based on the idea that tuberculosis was rarely eradicated, only 'arrested' in the common medical parlance. Something like the 'alcoholic' produced by late twentieth-century AA discourse, sanatorium treatment produced a permanent consumptive. And in some ways like the contemporary subject who learns to 'live with HIV/AIDS', the consumptive learned to live with the disease in the least harmful way, rather than to expect cure. Consumptives were to think of themselves as always potentially infective. If their symptoms disappeared, it was because the infection had been contained, rather than eradicated, and inattention to their open air regime could threaten a recurrence of symptoms at any time. Thus part of the re-education of the dangerous consumptive concerned the instalment of a constant vigilance about their conduct, as well as a recognition that they were always a potential danger to those around them if they did not remain vigilant. They were to be responsible about their habits of living both in their own interests and in the interests of the community. This was part of the bargain of their 'release', so to speak, from institutional isolation. The aim was to make certain modes of conduct and interaction entirely habitual, yet at the same time to instil a mentality of constant self-monitoring on the part of each consumptive. It was a case where, as Rose has written, 'the "soul" of the citizen has entered directly into political discourse and the practice of government'.[83]

Training in responsibility to others worked from complicated premises, some emotional and moral, and some drawing from social contract imperatives. Competing obligations and rights were one of the languages for the education of consumptives in civic responsibility. For example:

> The patient's conscience may be appealed to, but his self-interest will form the most effective leaver in dealing with him. He must be taught that even public carriers are subject to regulations ... and that while consumptives have legal and moral rights, the general public, too, have theirs.[84]

Along the same lines, when the invalid pension for tuberculosis was discussed by the Commonwealth Government in 1916, the possibility of making the pension conditional on the safe and hygienic conduct of the consumptive was considered. As it stood, the Invalid Pension Act did not require that the tuberculous pensioner recognise any 'duty to the community in return for the care the community gives him'. The committee recommended that the payment of pensions be conditional upon the observance by the patient of at least the more important of the precautions recognised as necessary, as, for example, 'sleeping by himself, using proper sputum cups, &c'.[85] Being responsible as a consumptive was about citizenship, as conscientious monitoring of one's own infectiveness and dangerousness, and as reciprocal obligations to the community.

The ambiguity in the open-air regime between treatment and prevention carried over into the way that consumptives were seen, and saw themselves. If successfully trained in the sanatorium, they were not only considered 'safe' in the community, but were also a means by which hygienic conduct was to be disseminated to the public. Consumptives successfully instructed in the open-air regime 'provide the whole community with conspicuous object lessons in hygiene'.[86] The anonymous 1907 consumptive wrote: 'The Sanatorium is an education centre, and from it light is expected to shine into dark places, every patient acting as a torch'.[87] In the community, the reformed consumptive was not only 'hygienically harmless' or 'safe' but was a positive object lesson for those around her, a kind of missionary who actively taught by example. Rather than spreading the disease, she was to spread a new way of living and being. And in this particular case she seems to have done so successfully. In the coda of the book she wrote that she and her family were healthy, that she was a total 'convert' to open-air living and that her family all slept outdoors religiously:

> Dickie and Rosemary sleep out doors in all weathers, and are rosy as apples. None of us caught colds this winter. Even the baby boy slept out; he is a fine three-year-old, so sturdy ... I think I am most inclined to dogmatise upon the subject of open-air sleeping. It is so exhilarating, so gloriously refreshing to sleep under the stars.[88]

As institutions, sanatoria were far more than places where the dangerous were isolated for public safety, although I think it is important to note that they *were* that, in the tradition of quarantine. They also produced reformed consumptives, whose hygienic conduct ideally not only prevented further infection but also produced advocates and exemplar of open-air living in the community. These were people who 'Sowed the Seeds of Good Health' in an increasingly eugenic formation of government, self-governance, population and hygiene.[89]

Conclusion

The schema 'from quarantine to the new public health' pursued by many historical sociologists is both generally correct, but nonetheless very general. What interests me about the case of tuberculosis and the sanatorium is that both *new* quarantining strategies for the confinement of the dangerous and, within quarantined space itself, new modes of training-into-healthiness came together. Tuberculosis management at the turn of the century represents a complicated simultaneity of public health measures and rationales involving isolation, rather than a progressive teleology of measures. Newly imagined as a communicable disease associated with urbanisation and industrialisation, tuberculosis management was re-conceptualised within longstanding spatial strategies of segregation. Those with symptoms, and importantly, those who were carriers, found themselves caught within a discourse of dangerousness, one manifestation of which was a modern form of quarantine: the sanatorium. But I have also discussed and suggested the nature of this institution as *modern*, as one which worked as much through 'voluntary' submission to isolation, as through coerced and legislated means of exclusion. And as comparison with subsequent chapters will show, this voluntariness was predicated on the freedoms of whiteness. Sanatoria represented both preventive isolation, thus protection of the general community, but also therapeutic isolation,

80 *Imperial Hygiene*

which took the form of a programme of re-training and disciplining into healthy and safe living.

The sanatorium then, was neither like the quarantine station nor like the island leper colony, but had elements of both. The policing of consumptives was sustained throughout the period. Or more correctly, *new* powers were created for the regulation and sometimes the compulsory isolation of those persistently represented as 'dangerous' and ungovernable. But unlike the 'dangerous' leper whose conduct and psychology, as we shall see, were of no particular interest to experts and authorities once they were confined or removed, the conduct and psychology of the consumptive was precisely the *project* of sanatorium 'treatment'. 'A sanatorium is not merely a hospital in the ordinary sense ... it is also a training school where patients are taught how they must live in order to overcome the disease'.[90] Far more refined examples of 'productive' power than the quarantine stations, the ideal sanatorium's disciplinary regime aimed to radically reform the tuberculous person, who voluntarily submitted themselves, their selves, for re-making. The oftentimes voluntary nature of their submission to institutional discipline and isolation, their complicity in the project of learning a new way of being, was itself evidence of new modes of governance at work. This was a place which aimed to enclose people spatially, and within that enclosure to cultivate healthy habits, indeed to produce responsible and civic-minded, therefore safe, consumptive subjects.

4
Leprosy: Segregation and Imperial Hygiene

In the nineteenth and twentieth centuries, leprosy was not exclusively a colonial issue,[1] but for most English-speaking governments, scientists, epidemiologists and public health officers, leprosy was thoroughly organised through, situated in, and productive of questions and imperatives of race relations and colonial rule.[2] It was, unlike tuberculosis or smallpox, often conceptualised as a 'tropical disease' even though it was a public health problem in decidedly untropical places such as New Brunswick, British Columbia or Robben Island in Cape Colony. Indeed the tropical medicine expert Patrick Manson wrote of 'a good many lepers in Iceland' but such exceptions did not stop him, or his colleagues in the field from comprehending leprosy as a disease of 'tropical and sub-tropical countries'.[3] Despite serious questioning over the mode of transmission of leprosy, systems of isolation were implemented in varying degrees of rigidity and in numerous imperial locations, especially from the late 1880s. This was not incidentally, but rather intimately, even causally related to the colonial context of much leprosy management: the non-whiteness of so many people with the disease.[4] As an 'imperial disease' the control of leprosy became entangled with spatial governance of indigenous people throughout the British Empire, with colonial laws as well as local rule regulating movement, contact and institutionalisation. But the nature of these exclusions and enclosures of space was not straightforward and was the object of considerable expert inquiry within the human sciences and the (often crossover) biomedical sciences. Leprosy was newly 'an imperial danger':[5] white people were contracting the disease. The disease itself, as well as anxiety about it, connected Empire through British migration, through the Chinese diaspora and through the circulation of goods. In its newfound imperial and colonial management, there was an imperative to differentiate just, British, modern systems of institutionalisation and/or segregation of lepers from premodern, 'medieval' systems and from the unjust exclusions of local, indigenous custom. The nature and powers of segregation, prevention and care often entered into much larger debates about 'modern methods', civilisation and compulsion as well as being inflected by Christian discourse of mission and salvation.[6]

While Michel Foucault used the premodern treatment of lepers to instantiate crude and unproductive powers of 'exile-enclosure', he could well have sustained his examination of leprosy management into the modern period.[7] In some colonial instances, and in response to the widespread problematisation of coerced exclusion, some leper colonies were (in principle) architectural and administrative utopias of governance through freedom, and through ideas of citizenship.[8] The Australian island leper colonies which I examine here, however, did not represent such an instance, being among the most rigid and isolated in the British Empire and Commonwealth. Created from the 1890s and in use in some cases until the 1970s, they represented an unremitting segregation, always racialised, but increasingly explicitly so as the twentieth century progressed. This was 'exile-enclosure' in the late modern world.

Leprosy management went on not only within the confines of the various leper colonies, the focus of most of the historiography, but also without, in the social domain. Capturing the complicated play of race, space and power in this instance, requires analysis of the social spaces either side of the *cordon sanitaire*. While race and disease were managed through crude measures of isolation and the imposition of rigid borders, this was not the only mode of governance at work. Leprosy management had social effects well beyond the shores of the island-lazarets, in large part because leprosy was such a deeply inscribed way of thinking about race in the Australian case, both internal relations between British-whites and Indigenous people, and external relations between British-whites and Asians, in particular the Chinese diaspora. In the final section of this chapter I discuss how in Australia, leprosy was understood by several influential experts as being transmitted sexually, in particular through sex between races. The 'health' management of leprosy became closely intertwined with the management of racial contact and conduct.[9] What I call 'racial *cordons sanitaries*' saw an almost complete conflation of race and health spatial management in instances outside the leper colonies altogether.

'An Imperial Danger': contagion and segregation

In the imperial nineteenth century, this chronic and minimally transmissible disease became the focus of much debate and policy on prevention: new statutes empowered authorities to detain people with leprosy, new spaces of isolation were governed both religiously and medically, international congresses debated cause, treatment and methods of prevention, new philanthropic enterprises and medical charities were established in England.[10] The reasons for this expansion of medical, legislative, religious and charitable concern about leprosy were in part the same reasons that tuberculosis, diphtheria or syphilis became newly (or differently) problematised – new expectations of government intervention in communicable disease, new techniques of diagnosis and notification, experiments in public health detention. But leprosy was also different: it had 'disappeared' from Europe and 'reappeared' in European colonies. Except for some morbidity around Norway, the disease was more or less unknown in Europe after the sixteenth century. The Norwegian exception increasingly tantalised European epidemiologists and clinicians in the nineteenth century as they pondered leprosy's hereditary or contagious nature,[11] but for many the great problem of leprosy was by then experienced in the colonised world. Moreover, it was not only that large numbers of people in India or the Straits Settlements or the West Indian colonies had leprosy, but that in many colonial contexts increasing numbers of whites were being diagnosed with the disease, contracting it, it seemed, from non-white people. Ironically, of course, while the fact that white people contracted leprosy in the colonial world drew it into a new frame of reference, leprosy continued to be racialised as a 'black disease' in places such as the Cape and the Australian colonies.[12]

From the 1860s, leprosy and its management began to be interrogated as a problem of Empire. Systems of control were compared across the vastly different colonial populations and legal systems, originally in an extensive Report of the Royal College of Physicians prepared for the Secretary of State for the Colonies in the early 1860s. This drew detailed information from medical officers in New Brunswick, Jamaica, Trinidad, Antigua, British Guiana, Mauritius, Madagascar, Hong Kong, Macau, New Zealand, New South Wales, Victoria, Tasmania, Ceylon, Bombay, Madras and Calcutta.[13] By the turn of the century, when responsibility for the 'health of the empire' became firmly entrenched alongside the institutionalisation of tropical medicine and through the administrative directions of Chamberlain, the control and even eradication of the 'imperial danger' of leprosy became part of the imperial mission.[14] 'With the expansion of England' wrote one sensationalising writer in 1884, '[leprosy] has been brought back to our very doors ... wherever the eye may rest throughout our vast Indian Empire, or the further-stretching limits of our colonial possessions, the dark cloud of leprosy is at this moment ... overshadowing the fairest spots of earth and the most fruitful territories of our commonwealth'.[15] Such literature sometimes spoke of the eradication of leprosy from the British Empire as if, as a governmentally linked entity, the Empire was also contiguous space which could erect *cordons sanitaires* around itself. But the danger was well within imperial space. Writers were anxious about the 'clean' of the Empire, and their indirect contact with the unclean: the Empire literally connected and circulated goods and matter, as well as people, within systems of constant communication and exchange. This was not a metaphoric fear, but one of actual contamination. For example, leper colonies and quarantine stations across the world had elaborate rituals for the disinfection of letters.[16] When Agnes Lambert described the extent of leprosy within the Empire in her series of articles published in *The Nineteenth Century*, she did so by constantly drawing attention to the exchange of goods between the colonies and England. The 'loathsome leper' is in 'productive Trinidad' for example, 'whence come our sugar, cocoa, molasses, rum, coffee ... and choicest West Indian Fruits'. Thus how the leper was segregated or treated in Trinidad was of direct concern for any English person. In such formulations, the English were not necessarily untouched by such contagion, 'tropical' or 'colonial' though it may be.

Not surprisingly, there was considerable medical discussion between epidemiologists, colonial medical officers, bacteriologists as well as new self-appointed 'leprologists'. As with so many diseases in the period, debate turned firstly on the aetiology of the disease, secondly and relatedly, on the most desirable and effective method of prevention. Less discussed was treatment, although as we shall see there were innovations in the early twentieth century and interwar years. Was leprosy contagious or hereditary? If contagious, what kinds of *cordons sanitaires* were most possible and effective, for which populations, and in which contexts? In the early 1860s when the Royal College collected and collated the reports on leprosy from the colonial medical officers, the meaning of contagion itself was actively debated and was certainly less than clear.[17] Yet determining a disease as contagious or not also determined (in theory) rationales for temporary quarantine or permanent segregation.[18]

In the case of leprosy, the question of its contagiousness was most important to settle because that would also settle the justification for confining or segregating lepers, both the numerous indigenous systems already in place across the Empire (and beyond) and methods of segregation which had been instituted in an ad hoc way through imperial government. Indeed the inquiry that led to the 1867 Royal College report was prompted by governors in the West Indies seeking guidance on the legitimacy and necessity of incarcerating lepers. The College asked colonial officers whether they considered leprosy to be transmitted as a contagious disease. This question was followed logically and importantly by a question on management: 'Are persons with leprosy permitted ... to communicate freely with the rest of the community? or is there any restriction imposed, or segregation enforced, in respect of them?' The College concluded strongly that the disease was not contagious 'in the ordinary sense of the term – i.e. communicated to healthy persons by direct contact with, or close proximity to, diseased persons.'[19] In finding against the contagiousness of leprosy, the Royal College also made forced confinement and segregation an unnecessary mode of management. Indeed the Report was sent to all colonial governments with the request that Governors attempt to abrogate any existing laws and customs for the compulsory segregation of lepers.[20]

Many of the local doctors, although not all, reported strongly against the contagiousness of leprosy. From Bermuda, the Physicians heard that 'during more than 35 years experience ... I have never been able to trace the disease to contagion or infection'. From Jamaica: 'I am certain it is in no way contagious'. From the Cape: 'I have never been able to trace the disease to contagion'. From Barbados: 'Without doubt it is hereditary ... I believe it cannot be communicated by direct contact, and is therefore not contagious'. From Calcutta: 'Never'. Indeed, while some of the colonial doctors did indicate that isolated cases seem to have been infected from another person, most of them located the contagion theory for leprosy as part of local and indigenous understandings. They were often insistent about this: 'I have never met a single instance'; 'the evidence against contagion ... is irrefragable [sic]'. This insistence should be read as a technique to distinguish themselves from popular and folk belief, especially in the West Indian colonies. The Antigua respondent, for example, said that 'the disease is considered contagious among the people of the colony generally; but I never have met with any case'. And the medical officer in St Lucia: 'It is commonly believed among the lower orders to be so; but the belief is confined to them'. In Jamaica there was a 'popular belief in contagion. But this erroneous opinion should be discouraged, as being unjust to the unfortunate sufferer'. And a respondent from Canton wrote that 'it is affirmed to be so by the Chinese, who regard it with horror. The law regards and treats it as a contagious disease'.[21]

In these early discussions of the return of leprosy and its constitution as an imperial problem in the 1860s, fear of contagion was often associated with a traditional or irrational folk belief. This was linked to imperial caution over local customs of exclusion, which of course varied widely, but were often taken as indications of uncivilised and unjust rule. In the way that the treatment of women was sometimes taken as a yardstick of a society's un/civilised rule, so the treatment of lepers were measures of indigenous and colonised societies. Leper exclusion was also linked with the irrational prejudices of the European middle ages. In Syria, if a leper would not go to a public asylum, one doctor reported, 'they are made to live in a cave or hut outside the village, where they remain in perpetual quarantine'.[22] In Cyprus a leper 'is torn from his family ... his goods are divided amongst his relatives, and he is banished from their presence for ever'.[23] British rule brought liberal fairness and replaced local prejudice with more civilised justice, according to many. 'Whatever Englishmen may feel about the acquisition of Cyprus', wrote Lambert, 'the leper at least is thankful that Cyprus has become a part of the British Empire'.[24]

Such commentary sought to distinguish just British management of lepers from traditional exclusions. Yet British asylums and hospitals in the colonies were also institutions and practices of isolation, even if the rationale was a protective and therapeutic one, rather than the cruder preventive or punitive rationale which characterised some traditional exclusions (and which certainly characterised the coerced segregation which was to emerge quite suddenly in the late 1880s and '90s in many parts of the Empire).[25] What really went on across the Empire, was not a 'release' of lepers from local prejudicial exclusion as commentators like Agnes Lambert would have it, but rather more the introduction of institutions of confinement, the asylum or the hospital, which were understood by the British as humane and therapeutic spaces, not compulsory, but desirable on the part of the inmate. The asylum in the colonial context could be turned to the purpose of most any 'problem population' – criminal, vagrant, medical, social, racial.[26] As 'hospital' it could be understood not as primarily segregative, but rather as primarily therapeutic and caring. 'Isolation' was partly in the tradition of institutional philanthropic care of those unable to care for themselves, a responsibility keenly felt in many colonies. Colonial leper confinement also took shape within and meaning from segregative religious practice: missions of course, especially in the colonial world, were often spatially organised and relied on some degree of segregation. A leper colony for indigenous colonised people could be thought of simply as another kind of religiously governed mission providing care and instruction. The leper colony at Molokai in Hawaii is the famous example here. British imperial institutions and religious spaces of confinement for lepers were thus part of a continuum of widespread practices of exclusion of the leper, both 'mission' and 'mandate' as Worboys has put it.[27]

But as the nineteenth century progressed, the meaning of institutionalisation and/or isolation of people with leprosy shifted broadly from being primarily religious, philanthropic or therapeutic to being primarily preventive and protective of the community at large. Sometimes the former slid into the latter on the same ground; that is, as the same institution. In the Bahamas in the 1860s there was no law to prevent lepers from moving freely, yet 'the colony has endeavoured to prevent it by establishing a lazaretto in conjunction with the asylum'.[28] And at the Cape: 'There is no law authorizing the deportation of any leper … the Government provides a very comfortable asylum for all lepers; but its insular position [on Robben Island] deters many'. The response from Ceylon was interesting as it anticipated the kind of voluntary separation which in the twentieth century many experts and authorities sought from people with leprosy: 'There is no legislative restriction … but there is a public asylum to which the poor and unfortunate sufferers voluntarily resort. Those who are well to do remain in their own houses and among their own families, but never freely mix themselves with the rest of the community'.[29] As the official desire for segregation increased, the geographic or architectural separation of an asylum came to be not an incidental aspect of institutionalisation, but the very purpose of official action: not a building which protected the suffering, or which housed the vagrant, but a *cordon sanitaire* separating the infected from the community. In New Brunswick, for example, Governor Gordon described the Lazaretto thus: 'The outer enclosure of the lazaretto consists of a grass field containing about three or four acres. Within these limits the lepers are now allowed to roam at will. Until lately they had been confined to the much narrower bounds of a smaller enclosure in the centre of the large one, and containing the buildings of the hospital itself.[30] The question here was: 'where are the lepers in space', not 'how can we relieve them'.

At the International Leprosy Congress in 1897 the Royal College's firm decision against the contagiousness of the disease and against the justification of exclusion was reversed with authority: 'Isolation is the best means of preventing the spread of the disease ... and a system of obligatory notification and of observation and isolation ... is recommended to all nations'.[31] By the time Robert Koch received his Nobel Prize in 1905, he was arguing strongly for strict isolation for chronic contagious diseases, linking tuberculosis and leprosy in this respect.[32]

Despite the attachment of mandatory social exclusions with popular prejudices, this was precisely the turn that much colonial law and practice took in the very late nineteenth century: spatial exclusion became much more rigid. This process drew partly from the celebrated discovery of the bacillus by Hansen in Norway and the introduction of confinement laws there in 1885, confinement which was, by contrast, often domestic and never rigidly enforced. Across the Empire there was a strong trend toward the spatial isolation of lepers brought about through new compulsory segregation laws: British Columbia in 1886, New South Wales in 1890, Natal in 1890, Queensland in 1892, Cape Colony in 1891, Ceylon in 1901, Canada in 1906.[33] In India there was ongoing official inquiry into the contagiousness of leprosy and the need for segregation. A Leprosy Commission reported in 1891 against the need for segregation and the Lepers Act of 1898 was a compromise piece of legislation which targeted vagrant or pauper lepers.[34] These new legislative powers sometimes created carceral spaces out of already existing institutions, and sometimes required entirely new enclosures of space. In both cases, 'lazarettos', 'leper colonies', 'leprosaria' were, initially at least, far more like the quarantine stations, than productive disciplinary institutions like the sanatorium. In some colonial contexts this was to change, in others, notably Australia and the Cape, it did not.[35]

It is no coincidence that the string of leprosy laws from the 1890s through the first decades of the twentieth century accompanied the explosion of Chinese exclusion acts and immigration restriction laws which I discuss in detail in Chapter 6. The association between leprosy and the Chinese diaspora was a strong one and China itself was understood to be a global reservoir of the disease. Leprosy was endemic in China: if experts offered numbers of lepers in other nations across the globe – even the 200,000 in India[36] – in China 'lepers are innumerable'.[37] Chinese migration, goldseeking or indentured labour, were understood to be the routes for the entry of leprosy into 'British' space – either the Empire imagined as contiguous territory around which *cordons sanitaires* of immigration restriction should be placed, or entry into Australia or Canada imagined as quarantined, defended and white nations.[38] Likewise, there was an intermittent practice of deportation beyond national or imperial lines, of Chinese lepers who did not 'belong' within Imperial or Commonwealth territory or systems of obligation. For example, the Australian government was impressed with the arrangement between the British Columbian and Chinese governments which saw the deportation of the Chinese from the leper colony on D'Arcy Island, between the mainland and Vancouver Island. The Australians thought that 'the one chance of eradication is deportation'.[39] Indeed, it was suggested to government in 1907 that Europeans in Australia with leprosy be deported to Java where there was evidently no restriction on their movement. A policy of removal outside the nation, instead of segregation within, was seen to have the dual advantage of better protection for the community, while being 'more humane' for the leper concerned.[40] Under this logic, white people lost civic status by virtue of their leprosy, in the same way that other people lost civic status (or never had it) by virtue of their race. The extension of existing colonial powers to repatriate or deport 'foreign' lepers, that is people foreign to the Empire, was strongly advocated by the British Empire Leprosy Relief Association.[41] Thus summarily removing lepers from the Empire was one way of dealing with this particular 'imperial danger'.

Despite ongoing uncertainty and dispute over cause, treatment and prevention, documented in detail by Worboys,[42] there were developments in the management of leprosy and segregation over the last half of the nineteenth century, and into the first decades of the twentieth. As we have seen, in the 1860s, leprosy was conceptualised by British experts in the main as *not* contagious, and therefore segregation was not recommended. Indeed exclusion then seemed to spring from popular and local beliefs. By the turn of the century there was a general (though still disputed) understanding of leprosy as contagious, with a hardening of existing segregative measures and the introduction of new laws and practices, including quarantining, deportation and immigration laws: medical segregation was instituted as a 'modern' preventive technique.

But there was another and subsequent twist in this history of socio-medical segregative powers and their meanings. While the need for modern total segregation had always provoked some question and protest, not least from those 'inside', around the 1920s one of the world's leading experts on leprosy, Britain's Sir Leonard Rogers embarked on another 'modernising' campaign with respect to leprosy and compulsory isolation. For Rogers, in the tradition of one strand of the anti-CD Act protest, compulsory confinement after notification was ineffective and counter-productive simply because it deterred the infected – especially the early-infected – from seeking medical help. Rogers and his British Empire Leprosy Relief Association believed they had a cure for leprosy, at least a treatment which minimised symptoms – injections of chaulmoogra oil and its derivatives. In his eyes, then, anything which put lepers beyond the reach of experts also denied them this treatment. For Rogers, like his counterparts in the 1860s, if for different reasons, compulsory isolation was considered a backward measure: 'The age-long and nearly universal custom of compulsory segregation ... formed the greatest obstacle to the adoption of the modern plan'.[43] As in 'Biblical times and in the Middle Ages' compulsory segregation meant 'imprisonment, usually for life, such as has never been used in any other chronic disease'.[44] Elsewhere, for Rogers, compulsory segregation was 'drastic and cruel', 'based on ignorance', 'largely unfounded'.[45] Not only the indigenous of the Empire, but many Englishmen and women themselves irrationally feared leprosy, according to Rogers. His personal mission was 'modern treatment'. Leprosy was contagious, but not highly so, and was often caught 'domestically' – what he called 'house infection' – through cohabitation over many years. As Worboys shows, Rogers and the Association wanted to see settlements or farm colonies established, on the model of farm colonies for the consumptive, where they could be instructed in hygienic living and be available for systematic treatment.[46] Most importantly, he wanted these to be entered into voluntarily. Rogers and the Association, then, did argue for institutionalisation, not for preventive reasons but so that the population of lepers could undergo the treatment methodically, and so that they could be monitored.

Across the Empire governments, public health experts, colonial administrators as well as enclosed lepers and their advocates debated the nature of the rationales for, and the powers by which lepers were segregated, as well as the kind of space and institution into which they were put. Leper enclosure *said things* about British rule and British justice. Like so many modern institutions it was generally not tolerable that lepers simply be put away, as it were. Or possibly more accurately, it was not tolerable that this be *seen* to be the case. Simply casting out in a move of exile-enclosure, no matter how great the danger, was understood to be a premodern and illiberal move, precisely what characterised many of the societies which the British had colonised. Rather, colonial institutions for lepers entered into the tradition of therapeutic/corrective institutions, which as I have discussed in the case of tuberculosis, were moving between the workhouse/asylum model to the farm colony or industrial settlement idea. By the interwar years in many instances, the *productivity* and participation of those confined or segregated was considered crucial both to their health and wellbeing, and to the modern and civilised nature of the institution or settlement. In particular, the more the lepers segregated themselves and participated voluntarily, the more successful the segregative and institutionalising move was understood to be by authorities. Consent was sought, or at best was manufactured as part of the process of segregation. Vaughan has shown this of several African colonies and Anderson has argued this of Filipino lepers at Culion, where isolation was about the productive training of lepers into civic subjectivity. What he calls 'the usual sad tale of stigmatization and segregation' did not apply at Culion, rather 'the leper colony became a laboratory of modern citizenship'.[47]

Foucault used the outcast leper as a model to illustrate crude sovereign powers of exile. By contrast, the enclosed plague town was his early example of disciplinary government. But he could well have sustained his analysis of leprosy into the late modern period. For as instances of disciplinary enclosures of space and developing sites of governmental power, they are exemplary. For example, one 'plague town' made all the more perfect because of its mechanisms of colonial rule was the tiny island of Nauru in the Pacific, beset by leprosy after 1911. The Island was German territory from 1888 to 1914, after which it came under Australian control as a mandated territory. Leprosy was unknown on the island before 1911, but thereafter it increased fairly rapidly. Its apparent virulence on the island was studied against theories of the evolution of racial susceptibility.[48] Assistance was sought directly from Rogers and the British Empire Leprosy Relief Association by the Administrator of the Mandated Territory and the Council of Chiefs.[49] Rogers, alive to the significance of the island's isolation was very keen, seeking to involve the Australian government. Through the Association, new diagnostic tests were made available and the latest version of chaulmoogra oil, hypnocarpus oil, was distributed and injected intravenously as well as intramuscularly, in addition to being taken orally.[50] The efficacy of this treatment for lepers in the earliest stages of the disease was noted. As a small island it was totally contained – no-one entered or left unnoticed in this place of suddenly intense surveillance and epidemiological observation. For all the effort which had gone into separating the populations on Molokai, Robben Island, D'Arcy Island in British Columbia or on the Australian leper islands, Nauru already offered itself as both isolated and segregated. Its isolation made it epidemiologically interesting within theories of racial susceptibility, as well as making the object-population containable and traceable for study. Additionally, the smallness of the population made knowledge of the colonised space near perfect for both government and epidemiological/clinical intervention: 'We have an excellent Card Index of the people' wrote the Administrator, 'commencing when a child is born and continued through life. By this means we are enabled to trace the life history of each Nauruan. We are able to exercise capable supervision of the movement of the people because our Districts are small and are controlled by Chiefs who have an intelligent knowledge of local government'.[51] Especially in its incorporation of local rule, this was perfect grafting of medical and colonial governance.

Constantly searching for representative models, the 'plague town' was less than perfect for Foucault because it was produced out of emergency situations, rather than being part of the everyday. The panopticon represented for him 'a mechanism of power reduced to its ideal form' in which, to quote Bentham himself 'morals [are] reformed – health preserved – industry invigorated – instruction diffused – public burthens lightened'.[52] But in many ways some of the leper colonies are far better models of utopian disciplinary governance than the Panopticon. Unlike the Panopticon, which drew architectural attention both to its powers of enclosure (its external walls) and to its place of surveillance (the central tower) the space of some leper colonies hid their own enclosing and policing powers, both architecturally and administratively. The report of one said:

> 'A Leper "Asylum" to use the old offensive name was simply a Sleepy Hollow where the afflicted folk passed a dull torpid existence. Things are different now. All our patients are eager to be well and they know that the more active they are the sooner they will be better. Laziness does not in any sense pay. Dichpali, therefore, is fast becoming a huge agricultural colony'.[53]

The Director of the Sungei Buloh leper settlement in the Malay States in the 1930s, to take a further example, took great pains to show in his report that authorities 'avoided any suggestion of control by coercion whether physical or by the attainment of a routine stupor among the patients'. Nor (apparently) did geography keep them there: 'The

Settlement is not on an island, detained patients could escape if they wanted to'. And the final measure of success, in his own terms, was the absence of boundary police: 'There are no outside guards or police, and the settlement is only a mile from the main road to the capital'. He said that this was achieved 'by placing the work and partial control of the institution in the hands of the patients themselves, and by the attempt to achieve a natural and fair atmosphere'.[54] What experts wanted here was governance through freedom, an alignment between the interests of the leper and the interests of administration. Freedom was constantly reiterated, precisely because of the act of enclosure itself. Many social theorists have argued that this kind of rule has come to characterise neoliberalism of the later twentieth and twenty-first centuries. The ideal subjectivity sought of the leper was not 'captive' but 'free': in Alan Petersen's words 'a sphere of freedom for subjects so that they are able to exercise a regulated autonomy'.[55] Authorities sought this cultivation of consent, civic identity and autonomy in many different kinds of institutions in the early twentieth century, but paradoxically it was often some bodily, racial or mental incapacity which put them into segregation, that is, *outside* the social body, in the first place. As we shall see, the rigid Australian laws and practices were a problem in this respect. In the British imperial context, leper exclusion was one site that both tested liberal rule, and, it was hoped, demonstrated the modern efficacy as well as humanity and justice of colonial government. But the changing nature of leper spaces captures the paradoxes of liberalism in the period and its 'boundaries of rule'. They show the tendency to demarcate and disenfrachise certain populations, examples of the 'exclusionary effects of liberal practices' which have so often accompanied the 'inclusionary pretensions of liberal theory'.[56]

Exile-enclosure: island isolation in Australia

The treatment of lepers in Australia ran directly counter to this (of course unfulfilled) imperial and liberal desire to institutionalise voluntarily and productively, in the Foucauldian sense. And it ran counter to the trend detailed in the last chapter of the voluntary presentation of the consumptive self-for-reform in the early twentieth century. By 1923, Rogers described the Australian system as being based on 'the most complete compulsory laws in the British Empire'.[57] Rogers' irritation with Australian governments and experts was extreme, as they insisted through the decades of the 1920s, '30s and '40s to implement mandatory isolation on islands for nearly all people with leprosy. Indeed Australian governments made these systems more not less rigid as the twentieth century progressed, in particular for Aboriginal people. 'When Will Australia Adopt Modern Prophylactic Measures Against Leprosy?' he demanded in the *Medical Journal of Australia* in 1930.[58]

In line with global concern with leprosy, new legislation appeared in the Australian colonies in the 1880s and '90s. The Victorian government amended the Public Health Act in 1888 to make leprosy compulsorily notifiable and to empower authorities to remove people to the quarantine station. Leprosy was made notifiable in Western Australia in 1889, and in South Australia in 1885 which then had power over the Northern Territory. In New South Wales, an 1890 Act made the disease notifiable and people with leprosy were to be detained at the Infectious Diseases Hospital – the Sanatory Camp of the 1881 epidemic – in a new lazaret. In Queensland, similar powers were introduced by amendment in 1891 and in 1892 a Leprosy Act was passed. Finally, the Commonwealth of Australia nominated leprosy as a quarantinable disease in its 1908 Quarantine Act to be discussed in the next chapter. In making a disease notifiable and quarantinable it was essentially being declared contagious. But many experts, it should be noted, still disputed this. Islands around the periphery of the continent came to be the standard sites of Australian leper colonies. A lazaret was built at Mud Island near Darwin in the Northern Territory, in the later 1880s. There were also new lazarets at Dayman and Friday Islands in the Torres Strait. When the first white person was diagnosed in Queensland he was isolated at Dunwich on Stradbroke Island near Brisbane.[59] Later, Peel Island, which had been a quarantine station was used, and housed together white people, the Chinese, South Sea Islanders, and Malays who had been on Friday Island, as well as Aboriginal people with the disease. After 1940, largely on the advice of Raphael Cilento, Chief Medical Officer and Chief Protector of Aborigines, Aboriginal people with leprosy were separated and located on another new lazaret on Fantome Island which remained in use until 1973.[60]

In all, the legal and government activity around leprosy in the Australian colonies was sudden and intense from the late 1880s. The amount of medical and public health effort as well as popular anxiety over the disease was out of all seeming proportion to the danger, on any criteria. Even those who knew it was contagious, also knew that leprosy spread with nothing like the virulence of smallpox, plague or even syphilis and tuberculosis. The numbers of people with leprosy in Australia was tiny compared with any other contagious disease, and with virtually any other nation or colony within the Empire. Director-General of Health J.H.L. Cumpston calculated 108 newly diagnosed cases between 1895 and 1900; 73 cases between 1910 and 1915; 93 between 1920 and 1925 (see Figure 4.1). The real public health problem for Cumpston and others was not the morbidity rate, but the changing racial distribution. As is evident from Cumpston's table of the 'nationality of recorded cases of leprosy', the disease was associated initially and strongly with the Chinese. During 1890–95, after the laws requiring notification were introduced and alongside the indentured labour systems, more South Sea Islanders – 'Kanakas' – were diagnosed. There was simultaneously a sudden increase in diagnosis and notification of 'Australians' and Aboriginal people. Thereafter, the Immigration Restriction Acts and the Deportation Acts to be discussed in Chapter 6, affected the number of Chinese and Islanders in the country, and the disease came to be primarily a problem of Aborigines and whites.

Leprosy was strongly linked to anti-Chinese agitation especially in Queensland.[61] But it was the fact that whites were being diagnosed that most turned leprosy into a government and epidemiological problem. Just why whites in Australia acquired the disease, but whites in

	Chinese	Kanakas	Aborigines	Other coloured aliens	Americans or other whites	Australians	Total
1860–70	30+	–	–	–	4	1	35
1870–75	15	–	–	–	–	3	18
1875–80	11	–	–	–	3	2	16
1880–85	18	–	–	–	2	2	22
1885–90	31	1	–	2	4	5	43
1890–95	27	10	7	3	5	19	71
1895–00	27	41	13	1	18	8	108
1900–05	15	43	7	5	14	19	103
1905–10	14	39	35	5	8	21	122
1910–15	6	14	22	4	10	17	73
1915–20	4	5	31	1	7	27	75
1920–25	3	5	54	–	9	22	93
	201	158	169	21	84	146	779

Figure 4.1 Nationality of recorded cases of leprosy in Australia
Source: Adapted from J.H.L. Cumpston, *Health and Disease in Australia*, introduced and edited by Milton Lewis, Australian Government Publishing Service, 1989, p. 209.

England did not, drove a monumental study funded by the London School of Hygiene and Tropical Medicine published in 1927 by Dr Cecil Cook, Chief Medical Officer, Quarantine Officer and Chief Protector of Aborigines in the Northern Territory.

> With the exception of Northern and North-Western Australia, the country to be considered is occupied solely by a European race of whom 98 percent are British. Leprosy, virtually unknown in the Mother Country except as an importation, is found, nevertheless, spreading amongst these Australian whites. What was the origin of the disease and why should a race, so rarely affected in its own country, become more subject to infection in Australia?[62]

For Cook, concern over the diagnosis of whites (mainly adult males) structured the possible questions to be asked in this epidemiology, as well as framing the imperatives of any preventive programme. In this view, Chinese or Aboriginal people with leprosy were in the main accompaniments to, conduits for, or causes of this problem. Another interwar leprosy expert E.H. Molesworth theorised about leprosy in terms of racial immunity, as was common for the study of many diseases and many investigators at the time. Molesworth explained the strange patterns in Australia through a theory which saw Aborigines as susceptible, and Europeans as immune except for 'the occasional throwback'. He also suggested that there was an elevation of the virulence of the bacillus through its 'passage through a member of a susceptible race'.[63]

The very small numbers of people with leprosy in Australia made it an intellectually exciting project for epidemiologists, and one which got them onto an international epidemiology circuit. This not only made total segregation logistically and financially possible – indeed the dream of eradication was realisable in Australia – but also facilitated a satisfyingly complete tracing of contacts. If historical and medical records were available, epidemiologists could be extremely precise in their case-by-case accumulation of data. Ashburton Thompson writing in 1897, a generation before Cook's *Epidemiology*, accounted for each of the 70 known cases in New South Wales to that date with extraordinary precision: '1859. A coloured West Indian observed by Dr Cox in a lunatic asylum. 1861. A Chinese admitted to St Vincent's Hospital ... 1868. White. The first of seven adult male whites was admitted to Sydney Hospital ... 1883. Five Chinese were admitted to a refuge in connection with the Coast Hospital (helpless; voluntary isolation) ...

1891. Four Chinese legally isolated' and so forth.[64] This highly individualised tracing permitted research into what Thompson called 'pedigree', the precise family history of lepers (see Figure 4.2). In the tradition of medical geography, place as well as race was for Thompson the crucial element in studying this disease. He focused considerable attention on the nationality and location of the various branches of each 'leprous' family – where they came from and where they had resided since. Cautious as always, Ashburton Thompson concluded little from his extensive study, while presenting a great deal of data. His strongest statement was the observation that white lepers born in Australia were all located north of the 35th parallel in New South Wales and Queensland. Further south in Victoria any white lepers had at some time travelled north, or had acquired the disease elsewhere.[65] This observation anticipated the West Australian 'Leper Line' of the 1940s, discussed below.

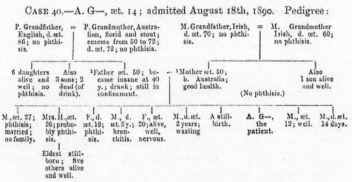

Figure 4.2 Leprosy as inheritable and inheritance as contagion: The Leper Pedigree
Source: J. Ashburton Thompson, 'A Contribution to the History of Leprosy in Australia', in Prize Essays on Leprosy, New Sydenham Society, London, 1897, p. 159. Courtesy, the State Reference Library, the State Library of New South Wales.

98 *Imperial Hygiene*

The powers of removal and detention of lepers were passed quickly by all the colonial governments, and unlike on-the-books powers for compulsory vaccination which were often not implemented, or even powers to detain people with tuberculosis or venereal disease which were implemented discriminately, nearly all people with leprosy in Australia were removed and detained, sometimes for their whole lives. With very few exceptions – usually wealthy whites who would undertake to isolate themselves in their own homes – all people with leprosy were isolated in one of the nation's quarantine stations or on the specially designated island leper colonies. The method of removal was implemented, as in the smallpox epidemics, usually through policing and penal systems. Such methods betrayed the heightened fear of contagion among locals and officials. One white man, a 'supposed leper', arrived at the Police Depot in Cooktown 'per train in a horse box ... as the Railway Department would not let him travel in a Railway Carriage'.[66] For Aboriginal people, forms of bodily removal and spatial coercion and confinement by authorities were problematically familiar, and it was this system of management which was utilised again for health reasons. In 1900 an Aboriginal woman was kept in a police yard, escaped and was apprehended again. The local sergeant wrote to the Commissioner of Police in Brisbane that 'the Gin was tied to a tree inside a galvanised iron stockade, 8 feet high situated in the Police yard'.[67] Generations later an Aboriginal man Nipper Tabagee told the psychiatrist and medical historian Ernest Hunter of police 'health patrols' in the Kimberley, Western Australia which continued until 1949: 'The policeman go round, one from Fitzroy Crossing, one from here [Derby], pick up those people from station and bring that truck. Chain them up first, bring them back to the leprosarium. After they out of their country you can take the chain off, frightened going to that other people's country'.[68]

People with considerable authority protested the powers of detention and removal. Ashburton Thompson, renowned for his work on the plague after the 1900 epidemic, was one of the earliest and most thoughtful critics, damning the 'extremely severe laws against the liberty of lepers which they have adopted, and at this moment enforce ... [They] add the remarkable hardship of imprisonment for life to the affliction of incurable disease'. He called this 'a flagrant infringement on personal liberty',[69] and drew on the recurring theme of 'medieval rule': 'The salient characteristics of those laws must seem ... to be a renascence in the nineteenth century of the product of medieval ignorance into medieval egoism.'[70] Recalling the arguments against segregation offered to the Royal College in the 1860s, the leprologist E.H. Molesworth wrote to a white woman on Peel Island: 'I personally do not believe in the need of segregation in leprosy, and I also believe that the persistence in such measures is a relic of an old and unjustified fear'.[71]

Other doctors questioned the grounds on which certain groups became suspect, and were subjected to blanket inspections. For example, in 1924 a white man was diagnosed and consequently all nearby Aboriginal people (not white people) were suspected of being infected. The diagnosing doctor was asked by the Commissioner of Public Health to bring all those Aborigines to Rockhampton in Queensland to be examined. He wrote in protest to the Home Secretary: 'I submit that this whole procedure is unnecessary and unreasonable. It is illogical for the blacks only to be examined – the whites are just as likely to be the source of infection if it come at all from any human being'. Such questioning of the basis of public health procedure paralleled earlier opposition to the Contagious Diseases Acts where women as prostitutes were detained and forcibly inspected, themselves illogically considered sole conduits for the disease, whereas the men they had sex with were not. 'The whites are left alone as they would not stand it', this doctor wrote in 1924.[72]

For Director-General of Health Cumpston, always alive to questions of government power, and constantly seeking education and consent, leprosy management was indeed a problem: 'is the general infectivity ... of such a high order as to justify absolute isolation?'. Yes, he concluded, but more absolute for some people than others. Through logic which was to increasingly govern policy on leprosy over the twentieth century, he distinguished the need for isolation according to other qualities of the leper: 'For Chinese, Kanakas and Aborigines, isolation under the strictest control is obviously all that can be considered. For Europeans who are indigent or feebleminded a similar control is necessary. The remainder of the Europeans present the great problem of leprosy administration'.[73] None of the Australian leprosy statutes explicitly distinguished between people on the basis of race, indigence or mental health. Yet such distinctions were constantly made on the ground, not only through decisions of a local police or health officer, but far more tellingly, through medical policy. What characterises Australian leprosy policy and management in the twentieth century is a trend towards more strongly and explicitly racialised practices of segregation. The discussion increasingly turned on questions of responsibility, on questions of capacity for responsible self-government.

For Cecil Cook it was Aboriginal sociality which justified continued isolation in their case. His perception of an Aboriginal person's 'careless and irresponsible habits … render it impossible to keep him under observation, or to submit himself to a course of treatment unless he is under restraint'.[74] Cook engaged directly with Leonard Rogers' call for less rigid isolation, arguing that since Aboriginal people could not be trusted to seek treatment, 'all lepers should be isolated in a lazaret without recourse to a bacteriological examination'.[75] And in 1934 it was stated that 'the effective control of leprosy and its eventual eradication are closely bound up with the supervision of the health of the Aboriginal population and other coloured peoples'.[76] Raphael Cilento, like Cook administratively responsible for both the management of Aboriginal people and for health, was also concerned about what he deemed irresponsible patterns of social behaviour as well as personal habits among Queensland Aboriginal communities.

> When the case of the aboriginal was investigated, the problem was seen to be infinitely complicated. The native habits of changing the name repeatedly further disguises relationships already masked by the haphazard use of the terms 'brother', 'father', 'cousin', 'uncle' etc. His complete dread of the white man's medicines, surgical possibilities, and hospitals (obvious in all areas, including those where leprosy is found most frequently) renders it utterly impossible to contemplate any system other than segregation for him.[77]

In 1950 the National Health and Medical Research Council (NH&MRC) presented a 'Standard Procedure in Respect of the Control of Leprosy' which categorised and nominated 'full-blooded natives' specifically as those who should be isolated.[78] And in 1956 at a special conference to review the 1950 policy and procedures, the NH&MRC reiterated this position. Despite what was by then a clear worldwide movement away from compulsory segregation, Australian health authorities argued that '[t]he time is not ripe in Australia for abandoning the present prophylactic system. It is in the interest alike of the patient and the general public that all cases of leprosy should be isolated and placed under treatment in special hospitals'.[79]

What characterised leper isolation in Australia was a marked disinterest in cultivating lepers' souls as it were, in the tradition of the sanatorium or even the workhouse or the penitentiary, and in distinct contrast to lepers spaces elsewhere. Rather, the primary objective of authorities, in a fairly unambiguous and sustained way, was segregation – exile-enclosure – hence the use of islands. Indeed at one point Cecil Cook refused to abandon the idea of an island leper colony in Darwin, even though it had been deemed unsuitable and inefficient, in particular lacking a fresh water supply. Unlike the intensity of the desire to build community and civic participation as in the Malayan case or the case of Culion in the Philippines, what the lepers did on their islands in Australia was never really part of the official project. If managers of the quarantine station and the infectious disease hospital in Sydney in 1881 had strictly separated men and women, Chinese and whites, steerage and first-class, infected and contacts, this heightened and imposed internal segregation was far less evident on the leper colonies. Officials and doctors sometimes complained about this,[80] but it seems that these kind of segregations were not strongly policed. On Peel Island, for example, there was an enclosed area for female white lepers but 'the sexes are practically at liberty to mix as they please despite the fence.'[81]

Most of the leper colonies had a manager on site, sometimes a husband and wife team in the manner of mid nineteenth century hospitals in Australia.[82] The managers saw to the distribution of goods and materials – bedding, clothing, materials to build huts, food – but beyond instilling a minimum order, the conduct of the lepers was certainly not their project. There was no sense in which they were responsible for the reform of lepers into hygienic and responsible citizens on anything like the sanatorium model. To some degree the managers spatially policed the inmates, but this was often as a form of punishment for insubordination or protest. In 1908, for example, one inmate wrote to the Home Secretary: 'I am writing to protest the treatment I am getting at the Lazarette to be keeped [sic] locked in a yard day and night not allowed to go out and talk to my fellow beings only to be left out for a short time each day in charge of a woman who was sent here as a nurse but who … is our jailer'.[83] The conduct management here was a crude order-control of a group of people whose protest was often against the act of isolation itself, who were criminalised precisely by this protest. 'What crime are we supposed to be guilty of', wrote one.[84] And the inmates at Dunwich in 1899 found it necessary to 'remind them [the Government] we are not prisoners criminals and that our lot under the kindest treatment is a hard one'.[85] Far from the cultivation of recreation and participation in their own administration of which the official at the Malayan settlement boasted, what Elkington then Chief Quarantine Officer wanted of the lepers in Queensland was the most basic level of order, produced on a punishment model: 'I do not advocate the use of drastic punishments against these unfortunate people, recalcitrant and difficult to deal with though they may be, but merely that support of lawful authority exercised for the common good and for the protection of the law-abiding, which is essential for the proper government of every community and institution'.[86]

Under this system, who the lepers 'were' did not in the main concern officials, so long as they kept some kind of order and did not turn into recalcitrant liberals. Ironically, however, the more rigid the isolation, the more heightened the sense of injustice on the part of not a few of the lepers and the growing number of advocates on the outside. In a way altogether different from the deliberate programming of civic identities at leper colonies elsewhere, precisely the totality of the exclusion in Australia produced a certain liberal or civic subjectivity. While the early correspondence between inmates of the leper colonies and authorities were largely about their conditions of detention, by the 1930s, more of their letters concerned detention itself. In 1939 for example the inmates of Peel Island wrote the Governor of Queensland requesting him to 'intercede on our pronlonged [sic] and wrongful detention'. They argued from the evidence of the Royal College of Physicians, poignant for its out-datedness: 'the condition commonly known as Leprosy is not transmitted by 'bacteria' of any kind from person to person, this confirming the report of the Royal College of Physicians'.[87] However, Raphael Cilento stuck to the isolation policy in the face of all international trend and what was to become the Peel Island Welfare Association took issue with him personally:

> The woeful persecution of the inmates their relatives and friends by Sir R. Cilento ... have been tolerated too long ... He scorns Dr Ernest Muir, all his authorities, the British Empire Leprosy Relief Association, together with all the world's expert full-time leprosy workers, their experiences, with the knowledge obtained thereby and put into text-books.[88]

In such cases it was largely the more literate white lepers who made on-paper bids for freedom, sometimes on their own behalf, sometimes on behalf of all the inmates. Aboriginal peoples' bids for freedom were more usually escape. These protests and communications ranged from supplicant petitions in the old English tradition, to invocation of law and science, to threats of lawlessness. One inmate wrote to the Minister for Health and Home Affairs to 'abolish segregation to see that we get justice, as the way we are being treated now is certainly not British justice and the conditions and restrictions at present in formed at Peel Island lazarette are a blot on civilisation'.[89] Failing to get justice though, placed the lepers outside obligations of the justice system, in their own eyes. In 1920 the Home Secretary was threatened:

> Having begged, and requested so often without avail, we now demand a visit from yourself to enable us to lay before you many grievances we are labouring under, failing which the consequences are yours. We would assure you this is no idle boast or threat ... It is said 'there is no law for Peel island', then naturally it cuts both ways? Whilst we wish to continue respectful and peaceful, we would ... again emphasise the patient's intention of doing something which perhaps may startle Queensland.[90]

As we have seen, however, it was not until the 1950s that releasing people with leprosy into the community was officially sanctioned in Australia, and then only for whites and bacteriologically negative 'half-castes'. A system that had begun in the 1890s within a strongly racialised fear of leprosy but which did not explicitly discriminate between lepers of different races, became over the twentieth century a system which explicitly segregated Aboriginal people but not whites.

Racial *cordons sanitaires*

The primary *cordon sanitaire* of leprosy management was the shore of the various island-leper colonies. Yet leprosy management was by no means limited to this quarantining measure, but rather involved spatial policing of racially identified individuals and groups in the social domain. In later chapters I show how, in the case of immigration restriction and quarantine powers, lines of hygiene were simultaneously racial lines. This was also the case in terms of internal health and race management.

Experts' opinions on the sociality of Aboriginal people were informed by, as well as themselves shaped the dominant culture which specifically excluded Aborigines from citizenship in Australia. If, as sociologists Petersen and Lupton have argued, 'the good citizen in the modern world is the "healthy" citizen',[91] perceptions about Aboriginal people's inability to perform 'health and hygiene' placed them outside the citizenry. Conversely, though, in that Aboriginal people were already outside the citizenry, questions about the legality and liberality of their compulsory isolation were to a considerable extent (although not totally) deflected. As many theorists have discussed, rights and freedoms of classical liberalism were bestowed not on all, but on those who were deemed responsible and capable of bearing the freedoms of citizenship. In concert with the diagnostic and classificatory powers of the human sciences, certain people have been deemed incapable of bearing these freedoms and responsibilities. These groups 'are thus liable to a range of disciplinary, sovereign and other interventions, including ones that we might recognize as "social" ', writes Mitchell Dean. Especially in the late nineteenth and early twentieth centuries, experts refined distinctions between the governable and the ungovernable, the employable and the unemployable. Liberal rule, he continues, 'is completely consistent ... with authoritarian rule of colonial societies'.[92] In Chapter 6 I discuss all kinds of eugenic distinctions made within the white population of Australia, as well as within the aspiring migrant population from Britain. But in the case of Aboriginal lepers this was a two-fold removal from the body politic. Right through this period, arguments about incapacity for responsible civic participation and self-governance placed Aboriginal people outside the citizenry of the social body, and inside demarcated missions and reserves. Infected, as lepers, they were literally and civically distanced again, not only from the social body of Australia but from family, land and community. As historian Suzanne Saunders has shown, for many Aboriginal inmates the coercion and forced containment in lazarets was an extension of the reserve systems whereby movement of Aboriginal people was limited, and in which people and families were already forcibly removed from one another.[93] Next to the exclusion from the franchise, the denial of freedom of movement was the clearest indicator of the civic status of Aboriginal people.[94]

In some ways, the reserves also functioned as policed quarantined spaces. Cilento certainly approached them thus, and in the 1930s and '40s he researched the transmission of leprosy in several largely closed Aboriginal communities, as part of a general increase in research interest in health and Aboriginal communities funded by the Commonwealth's NH&MRC.[95] Conversely, the spatial segregation of health institutions sometimes explicitly replaced the spatial segregation of the reserve system. This was the case with respect to the management of tuberculosis in Aboriginal people in the Northern Territory, for example. In the 1950s the establishment of sanatoria specifically for Aboriginal people with tuberculosis was discussed in the Northern Territory. Prior to that time, tuberculosis was managed within the missions, later reserves: isolation of the infected, exclusion from school and church and so forth, occurred within the already segregated racial space.[96] By 1953, just when antibiotic treatment was rendering preventive-therapeutic isolation in the sanatorium for the white populations of Australia irrelevant, the need to segregate Aboriginal people with tuberculosis returned under the logic of dangerousness and security. The diagnostic Mantoux test had been carried out through the Northern Territory in the early 1950s, revealing much higher incidence than previously thought.[97] It was now considered essential to 'segregate' and establish sanatoria at Darwin and Alice Springs, 'not only in the interests of the health and welfare of the Aborigines themselves, but as a measure of security and protection of the health of the European community'. This was directly related to the (minimal) reduction in spatial policing on the reserves which was taking place: 'the Aborigine presents a direct chain of infection from infected areas to the European community. With the relaxation of restrictive control of Aborigines they are moving about more freely and in greater numbers', wrote Cook in 1953.[98]

For Cook and Cilento in the interwar years, infectious disease and other forms of ill health were often produced by what they saw as illegitimate movement and intermingling of racial groups: the migration of people from their proper place to an improper place. Cook suggested the containment of 'pure' Aboriginal groups as permanently separate: 'In the virgin country of North Kimberley where the natives continue in their pristine state, the disease [of leprosy] is quite unknown'.[99] Conversely, the health of a community could be achieved by preventing such 'illegitimate' movement in the first place, by putting people back where, in his view, they belonged. This had been partially achieved by Queensland and Commonwealth immigration restriction acts, as well as the deportation acts. In his major work *The White Man in the Tropics*, Cilento warned readers about 'the repeated emphasis history places on purity of race' as a way of maintaining the health of a community.[100] Racial purity was a public health issue, indeed a preventive health strategy. The pressing issue for Cilento and for his understanding of tropical medicine as a way of purifying and populating the north was racial integrity, both Aboriginal and white. This objective of racial integrity involved a sense of spatial containment, at the very least the control and monitoring of movement of Aboriginal people. And it involved policing of heterogenous sexual contact, both within the nation as we shall see, and without. Indeed the Australasian Medical Congress resolved in 1920 that steps should be taken 'to promote an understanding with neighbouring civilized races, other than white, whereby the mutual advantage of the avoidance of hybridism and the perpetuation of pure races be made the basis of inter-national agreement'.[101]

The system of reserves to contain Aboriginal people in this period should be taken into account in arguing for the continuity of practices either side of the *cordon sanitaire* of the lazarets. Leprosy-specific segregating practices were considered and, in some cases, implemented on the mainland. Indeed Cook went so far as to consider a blanket racial segregation:

> Segregation of the race, it would appear, would be an efficient prophylactic measure ... Nor is it entirely impracticable. The enforcement of the existing prohibition against aborigines entering certain areas occupied by coloured aliens in Western Australia and the Northern Territory has not been seriously attempted. Even an abortive effort would not be without its advantages where a disease like leprosy is concerned.[102]

Even for those epidemiologists who argued against compulsory isolation – against the system of island-lazarets – racial segregation in the general community was nonetheless to be pursued as a public health measure. For example, E.H. Molesworth was a leprologist keenly active in the movement to abandon isolation. Yet in a heated debate with Cook in 1927 he argued for the importance of 'prevention of contact on the part of whites ... with aboriginals and Asiatics'.[103] Moreover, Molesworth conceptualised the advancing European frontier as a macabre, indeed deadly, public health solution. He thought that the colonisation of the north in the form of 'settled' white community would eventually bring about the disappearance of Aborigines altogether. So, while Aborigines were indeed considered the 'cause' of leprosy, 'with the rapid dying out of the aboriginals as a result of infection with tuberculosis, syphilis and other diseases and as the line of settlement advances', Molesworth argued, 'this problem will probably resolve itself.'[104]

In Western Australia the spatial projects of health and race became almost indistinguishable. The government, through a recommendation of its Health Department, passed the Native Administration Amendment Act, 1941 'in order that the spread of leprosy within the State my be limited'. It nominated a 'boundary line' which became known as the 'leper line' – the twentieth parallel of south latitude. It stipulated that 'no native who ... is living north of the boundary line, or who at any time thereafter shall have passed to that part of the State north of the boundary line, shall pass to any part of the State south of the boundary line'.[105] Permits were granted for nominated exceptional circumstances: firstly, the need for medical treatment, secondly the need to appear at a legal proceeding and thirdly, if the person was driving stock south with a 'white Boss'. In such cases, Aboriginal people could stay in the south no longer than three months, could be required to return north at any time, were to submit to medical examination, and were forbidden contact 'with other natives living south of the boundary line'.[106] Aborigines – not people with leprosy – who lived north of the twentieth parallel were forbidden to travel south, a restriction in place, if not enforced, until 1963. In these instances the spatial management of health and race had become the same project: these were truly racial *cordons sanitaires*.

Interior frontiers: sexuality, contact and race

The rigidity of segregation in Australia, and the extension of leprosy management well into the domain of colonial race management needs to be understood in the context of the peculiar theories of transmission in circulation at the time. Cook and others considered it to be transmitted sexually, more precisely through sex between races. Ann Laura Stoler has written that in many colonial racial economies 'cultural hybridities were seen as subversive and subversion was contagious.'[107] Sex between races had not always been culturally difficult, but by the 'degenerating' and eugenic early twentieth century, it was. It brought 'ill-health and sinister influences ... sources of contagion and loss of the (white) self'.[108] In Australia leprosy was a heightened hybridising danger, for it was not only a highly stigmatised contagion, but one associated with sex and with miscegenation.

The theory that leprosy was sexually transmitted existed as a minor strand in international medical literature, often working in an inferential way through a linking with syphilis.[109] Harriet Deacon has shown how there was a strong belief in sex as a mode of transmission in the Cape. Indeed on Robben Island, segregation by sex was more rigidly imposed than segregation by race.[110] In the 1860s the Royal College of Physicians had asked the colonial medical officers: 'Does the disease seem to be transmissible by sexual intercourse?' Nearly all the respondents answered negatively, but it is clear from the question that there was already an association.[111] In Australia, there was a larger interest in the sexual transmission of leprosy than seems to have been the case in other colonial and national contexts. At the 1884 Sanitary Conference in Sydney, for example, the delegate from Western Australia confidently stated that leprosy was spread by 'the prevalence of prostitution of white women to Chinese'. And a Queensland delegate argued that 'we have never had the disease amongst the aboriginals in Queensland ... Simply because the black women will not cohabit with the Chinese'.[112] The strong connection in white Australian popular and political culture between Chineseness, leprosy and national invasion, which was not uncommonly represented as sexual invasion of virginal white womanhood, may well have been the fertile cultural ground from which these medical/epidemiological links were made.[113]

Cook offered the theory of the sexual transmission of the disease thus: Chinese and Pacific Islander men, infected elsewhere, entered the Australian colonies as immigrants or as indentured labourers in the nineteenth century. They had sex with Aboriginal women who later had sex with Aboriginal men and with white men. For Cook the conduit for the spread of leprosy was Aboriginal women. There was no room in his theory for the possibility of sex between men or sex between 'coloured aliens' and white women. He described what he thought was the key to the whole question of leprosy in Australia:

> The matter of aboriginal gins is much more important, since the alien deprived of the society of women of their own kind, and unable, except in very rare instances, to overcome the racial prejudices of the white women, fell back for conjugal relationship upon the salacious aboriginal. In this way the races came into the most intimate contact ... herein lay the danger to the white ... Although the whites did not become directly associated with the Chinese and kanakas, there was ... a definite link between the two races by means of which the diseases of the latter could be transmitted to the former.[114]

The causative condition in this logic hinged on sex and race demographics, 'the lack of an adequate white female population' and the 'presence of Aboriginal women'.[115] Accordingly, Cook concluded that a 'Chinese or South Sea Islander leper is, generally speaking, only to be considered as constituting a menace to the white population where there is (i) a considerable Aboriginal population, and (ii) a scarcity of white women'.[116] Thus, for Cook, the infection of the white population could only happen 'per medium of the aboriginal'.[117]

Cook was involved in the business of sex and race from several perspectives: as infectious disease control; as part of interwar 'race science'; and as policy on 'the Aboriginal problem'. His expertise perfectly fitted him for his powerful dual position of Protector of Aborigines and Chief Medical Officer, because both of these domains concerned sex, race and contact. When he embarked upon his research for *Epidemiology of Leprosy* he had just completed a study of Aboriginal people and venereal disease.[118] More importantly, he was a central player in Aboriginal policy in the 1920s and 30s which saw a general shift from policies of separation and containment, to policies of assimilation: half-caste children were to be brought into the white Australian social body.[119] The new policy of assimilation had begun about 1920 and was agreed upon in theory by all Australian states at a 1937 conference. One version of the assimilation program worked through sanctioned programmes of sex and marriage between half-caste women and working-class white men.[120]

At the 1937 Conference, Cook saw the Northern Territory Aboriginal population under his nominal control as both piteously weakened and as dangerous. If left alone, he said, 'the aborigines would probably be extinct in Australia within 50 years. Most of the aboriginal women would become sterilised by gonorrhoea at an early age; many would die of disease, and some of starvation'. However, he said to his colleagues, 'if aborigines are protected physically and morally, before long there will be in the Northern Territory ... a black race numbering about 19,000 and multiplying at a rate far in excess of the whites ... their numbers will increase until they menace our security'. On the face of it, Cook was arguing that Aboriginal women should be left untreated.[121] In the same way that 'full-blood' and 'half-caste' lepers were to be distinguished in their confinement with respect to leprosy, Cook proposed a kind of sliding scale from assimilation to segregation for different Aboriginal people in the programme of official hybridisation. The 'coloured girls' were to be made 'acceptable to the whites' (that is, white men), the 'semi-civilized' were to be kept under 'benevolent supervision' and the 'uncivilized native' was to be kept inside 'inviolable reserves'.[122] Cook's work and ideas show how both race and health management worked within similar spatial systems: sometimes the aim was segregation, purity, quarantine and isolation, sometimes the aim was a *managed* contact, assimilation, merging, integration and hybridisation. In both cases, health and race were governed in the first instance spatially, in the second instance through the regulation of sexuality. This all came together for Cook as the problem of leprosy.

In studying the 'slow' disease of leprosy, connections and contacts were traced epidemiologically into the sometimes distant past rather than across the present as was the case with many other more virulent diseases. Leprosy in a white man in 1930 was traced through only several intermediaries to leprosy in a Chinese man in 1880. As Cook saw it, the contamination of the nation from outside in the past had left its traces very clearly in the internal present. The Chinese had been excluded from the nation but in some ways lingered as a threat. Indeed the question of leprosy can be seen as one of contagion between men of different racial cultures. In this way, the contagion anxiety was not only about unmanaged sexual relations between white men and Aboriginal women, but also about a mediated sexual contact between white and Chinese *men* over time. 'Although the whites did not become directly associated with the Chinese and kanakas, there was ... a definite link between the two races by means of which the diseases of the latter could be transmitted to the former'.[123] This was about loss of the white male self in a process of connections with unknown others. It was about miscegenation, but it was also about a kind of sexual contact/contagion, albeit at one remove, between white men and Chinese and Pacific Islander men. From Cook's perspective, 'their' diseases became 'ours' through illegitimate sex.

As we have seen, for race and health managers like Cilento and Cook, public health could be attained by securing boundaries between racial groups, by creating a newly permanent and stable social system in 'frontier' and tropical Australia: contact between races was prevented or discouraged except under the sanctioned politics of assimilation. One manifestation of this social plan was the encouragement of 'healthy' numbers of white women to the tropics, women who were influential in normalising and settling the problematic zone as racially and sexually stable. This interest in the numbers of white women in the north worked at several material and symbolic levels. Their sexual availability to white men supposedly resulted in a reduction in inter-racial sex. The familial and domestic cultures which they were meant to introduce represented 'settlement' and permanence in the tropics. And, as Anne McClintock has argued of women in South Africa and elsewhere, white women were markers of the nation at its precarious borders. In many national cultures, white women often symbolised the purity of an imagined community.[124] Rather than being imagined as the vulnerable white virgins open to sexual attack of outsiders they would be solid and strong mothers and wives.

This was part of the cultivation of the white self in tropical Australia, the precariousness of which I detail in Chapter 6 – identity and difference constructed and performed in terms of conduct and moral systems. The tropics needed securing as white, and this process of 'settling' in the north involved not only being there, as it were, but feeling settled: the white self needed cultivating.[125] This process might be thought of as the development of 'interior frontiers', borrowing from the theorist of nationalism Etienne Balibar. Analysing the early nineteenth century political philosopher Fichte's work, Balibar writes that an internal border or interior frontier is that which constitutes a community through internalised individual identity; for Fichte the borders of a 'spontaneous linguistic community', for Balibar 'the inner nation, the invisible nation of minds'.[126] Or, as Ann Laura Stoler writes: internal borders mark 'the moral predicates by which a subject retains his or her national identity despite location outside the national frontier and despite heterogeneity within it'.[127] Purity and integrity of an imagined national community can be threatened externally or internally. If the external frontier of the nation was secured by the Quarantine Act and the Immigration Restriction Act, the interior frontier of whites especially in the racially precarious North needed securing through other measures.

Cook wrote about the town of Derby in Western Australia as the type of healthy community to be developed: 'In Derby, being a permanent European settlement, the sexes amongst Europeans are comparably represented, and apart from that degree of association contingent upon domestic service, there is no intimacy or fraternization between the races'.[128] The public policy of encouraging white women to the north involved a range of 'civilising' and 'domesticating' meanings and processes. White tropical conduct was to be effected by women, and to be produced in and symbolised by, domestic arrangements. An industry in instruction in Australian tropical domesticity flourished between the wars, and it governed habit, conduct and attitude as well as management of the domestic environment. This included the minutiae of daily conduct; diet, exercise, clothing, literature to be read, leisures to be pursued, timetables for daily routine, as well as 'Attitude towards Native Assistants'.[129] The authority of this instruction moved across several domains of the human and social sciences, from psychology ('the problem of 'tropical neurasthenia') to 'tropical hygiene' (personal and domestic practices).[130] 'Settlement', Cilento said at one point, 'represents to me the Frontier Legion in that army of occupation that is seeking to make and keep Australia white'.[131] This referred both to the external threat that populated nations (Japan, China) would seek to use Australia's 'under-exploited' and 'empty' lands, but also to the internal threat of heterogeneity through social and sexual merging: the spread of leprosy to the white community was an object lesson. In this instance, as with so many hygienic imperatives, technologies of the self and technologies of the nation and of colonisation were the same. White Australia was only achievable through what Cilento called at one point 'personal prophylaxis', the perfect expression of individualising and totalising governmental power.[132]

Hygienic white women were understood, indeed simply assumed, to effect a reduction in sexual contact between white men and Aboriginal women. Cook based all kinds of conclusions on this assumption. For example, studying leprosy in the Queensland town of Bundaberg he concluded that the considerable numbers of European women indicated 'a degree of civilization and refinement in the community rendering combo-ism (cohabitation between white males and aboriginal females) highly improbable, even where there was a sufficiently numerous aboriginal population to encourage it'.[133] Race management and race science in Australia in this period were a mass of internal contradictions and inconsistencies and Cook's own ideas exemplify this.[134] On the one hand the assimilation program aimed to make 'the coloured girls acceptable as whites ... the female can be accepted as the wife of a white man'.[135] This required a very clear demarcation of the half-caste 'coloured' girl. On the other hand, and with respect to health and leprosy he suggested that a 'natural aversion' between European and Aboriginal needed 'fostering'. This would result from the presence of white women:

> As to the prevention of association between European and aboriginal, it is to be feared legislative enactments will be unavailing. On the other hand there exists in the white a natural aversion to these practices, which is only overcome after prolonged familiarity with degrading conditions and suppression of the sexual instinct. The fostering of the natural antipathy and the encouragement of female immigration such as will inevitably follow the development of the primitive regions where these conditions at present prevail, will do far more to segregate the races than a tome of prohibitive Statutes.[136]

This report announced a separation of races secured not coercively or through regulatory agencies, but rather through internalised lines of hygiene and as modes of gendered conduct and raced interior frontiers.

The presence of white women was perceived to socially separate the white social body, first from the Aboriginal community, and second from past invasion by Chinese men. Although this report suggested that the radical and forced limitation of movement of Aboriginal people was possible and desirable, the forced limitation of the sexual practices of white men was understood to be less so.[137] Rather, their desires needed to be manufactured (away from Aboriginal women, towards half-caste women or white women). In this way, the *cordon sanitaire* between races was never clear and straightforward but constantly reinvented.

While apparently effecting a reduction in sexual contact between the races, white women were perceived to be able to properly manage domestic and social contacts between Aboriginal people and the white world in their supervision of domestic help. All this was part of the cultivation of conduct specifically as white; the development of interior frontiers through an understanding of conduct which began with the domestic and the familial. It can be thought about as the introduction of a 'proper' private sphere to the north, intended to displace the illegitimate private, sexual conduct of white men and Aboriginal women which had ostensibly resulted in contagion and ill-health. This was a private which was performed in, and symbolised by, domesticity and family. In Stoler's words this linking of 'domestic arrangements to the public order, family to the state' was imperial/national biopolitics at work.[138] This was, indeed, imperial hygiene – 'colonisation by the known laws of cleanliness rather than by military force'.[139]

Conclusion

Like the management of smallpox and tuberculosis, leprosy drew into expert consideration movement and enclosures of space, contact between humans and its regulation. These were questions of citizenship. Here again, public health was about what and who circulated dangerously, and how legitimately and effectively that circulation could be controlled, eliminated or rendered safe. In the case of leprosy this was intensified and amplified because the disease also drew sex and race relations into official scrutiny. This joint management of race, health and sex was undertaken through sets of spatialised practices, involving boundaries, separation, quarantine, isolation and protection on the one hand, and anxiety about and regulation of contact, contagion, integration and assimilation on the other. Any contagion implied illegitimate contact, a transgression of a line of hygiene in one way or another, but in the colonial Australian context these were often racial *cordons sanitaires*. In such ways, racial lines and lines of hygiene were often indistinguishable in the boundaries of rule which governed both white and indigenous communities, and which constituted racial difference.

5
Quarantine: Imagining the Geo-body of a Nation

The long history of the *cordon sanitaire* and of quarantine practice is tightly bound up with the development of administrative government. The capacity to detain ships, goods and people from elsewhere, in the interests of one's own city, community or nation both presumed and tightened governmental authorities over commerce, health, and movement: over exchange and circulation. These technologies of government have been centrally concerned with the significance of population and population health to modern states and especially nation-states as they emerged over the nineteenth century. Public health and nationalism are both modern projects connected with the complex emergence of political economy and with the development of liberal democracy and concepts of citizenship. In this chapter I enlarge my focus on *cordons sanitaires* from analysis of the borders of bodies and urban spaces, to analysis of the connection between national borders and quarantine lines. Nations in the modern period always required mapped boundaries: they needed to be imagined and enforced as 'geo-bodies'. In *Siam Mapped*, historian Winichakul Thongchai writes: 'Territoriality involves three basic human behaviours: a form of classification by area, a form of communication by boundary, and an attempt at enforcing ... The geo-body of a nation is a man-made territorial definition which creates effects – by classifying, communicating, and enforcement – on people, things, and relationships'.[1] The explicit turning into discourse of geographic boundaries as well as the enforcement of these lines on-the-ground – their representation and administration – are part of what created spaces as nations, and nations as 'geo-bodies'. The national boundary came to be meaningful as a site of commercial regulation and customs, as a site of medico-legal border control and quarantine and, as I discuss in the next chapter, as a closely related site of immigration restriction.

Many histories of medicine and public health are framed in and by a national context, but not many explore how public health management itself sometimes shapes and informs national identities.[2] My suggestion in this chapter is that public health, in particular maritime quarantine measures, enabled a particular geographic imagining of Australia. 'Australia' itself was created at the beginning of the twentieth century. In 1901 six colonies of the British Empire were brought together into a new island-nation: the Commonwealth of Australia. A Commonwealth government thereafter overarched the state governments (the old colonies) and local governments throughout the new country. The 1901 Australian Constitution nominated one major public health power to be granted to the new Commonwealth Government: the power of quarantine. In this way, quarantine was (and still is) literally and constitutionally central to the knowledge/-discipline and the institutional bureaucracy of national public health. A national maritime quarantine 'line' was created, which *was* the border of the new nation. What type of imagining of 'Australia' was enabled, even required, by the idea and practices of quarantine? How was a language of biomedicine – epidemic, contagion, immunity, hygiene – tied up with a language of defence of nation – resistance, protection, invasion, immigration? Part of the effect of quarantine was the imagining of Australia as an island-nation, in which 'island' stood for 'purity' but also therefore vulnerability to invasion by infectious disease. The maritime quarantine line was seen to secure the nation at its border: it was a major 'measure of defence at the frontier'.[3] I also argue, then, that quarantine was culturally and imaginatively central to an early twentieth century nationalism in that it drew attention to the thresholds of the nation, one effect of which was the production of Australia as a 'geo-body'.

It is also significant that the period in which quarantine was being secured as the primary Commonwealth health strategy, it was being tied to racialised practices of immigration restriction which I discuss in detail in the next chapter. This period was one in which a new citizenry was being deliberately shaped, and in which the health of population was being articulated as a problem of national government. The constitution of the Australian citizen was under discussion. The layers of meaning of 'constitution' are significant here: who was to be 'citizen' or 'subject' in the Australian Constitution?; how was the Australian citizen to be constituted – made, educated, created?; and how to ensure the most robust constitution *for* each citizen – their health, bodily vitality, capacity for reproduction, powers of resistance to disease? In the contemporary West, the good citizen is the healthy citizen,[4] and this early twentieth century moment was one in which all kinds of agencies and expert knowledges disseminated desire for health, the moment when the state became increasingly responsible for health and welfare. This is true of other western nations, but what I think is most interesting about the Australian context, is that the early twentieth century was also the moment of self-conscious nation-forming and citizen-making, processes which remained by nature and in intention, deeply colonising with respect to the Indigenous population.

Quarantine and nationalism

Over the last half of the nineteenth century many public health and increasingly microbiological experts participated in important international sanitary conferences, in which 'sanitary' meant measures for preventing infectious disease: mainly quarantine, but also vaccination and to some extent 'internal sanitation' measures such as disinfection, urban sanitary reform, housing sanitation. These were prompted largely by anxieties about cholera epidemics in Europe and their source, later plague and with the entry of the United States, yellow fever.[5] Movement of people, particularly between India, the Orient and Europe, and within Europe itself, was heavily scrutinised. At the 1866 Constantinople Sanitary Conference the annual pilgrimage of Muslims from India to Mecca was isolated as the major cause of the spread of cholera, and especially after the opening of the Suez Canal in 1869, newly stringent quarantining and screening measures were introduced, at Indian ports, by the Egyptian Board of Health at the Canal, and in European ports. Various national and colonial authorities sought (not always successfully) to co-ordinate and standardise methods of infectious disease prevention, in particular quarantine measures where co-operation between governments had long been understood as imperative.[6] But the extent to which they could co-operate was always limited by different understandings of how much intervention was necessary to control the spread of disease, how much was desirable to minimise disruption to trade and commerce, and especially with respect to the post-1857 British policy of minimal intervention into Indian indigenous custom, how much intervention was possible for the British or Indian governments to impose without risking another rebellion.[7]

Within Europe and along global routes between west and east, blanket quarantine was imposed in certain years after a port was declared 'infected'. By the turn of the century, and largely as a result of the series of sanitary conferences, strict maritime quarantine as a public health strategy was being loosened, even abandoned as medically ineffective.[8] Strict quarantine (which as we shall see continued for some time in Australia) assessed risk and the need for quarantine primarily on knowledge of the state of health of the port from which the ship had sailed, and only secondarily on the physical state of the passengers. The 'International Bill of Health' for example, which was devised at an 1881 International Sanitary Conference in Washington DC, was a bill of health of and for the *vessel*, not individual passengers, based on the presence or absence of epidemic disease at its originating port.[9] Under this system (already seen by many to be outdated by 1881) the presence of cholera, for example, at the port of embarkation would require a period of quarantine of the ship, its goods and its passengers at port of disembarkation, whether or not there were any symptomatic passengers. Partly as commercial interests opposed strict quarantine, and partly as knowledge of incubation periods and modes of transmission changed, 'rational' or 'limited' quarantine developed whereby the need for detention was qualified to some degree by inspection of the ship and its passengers at port of entry. Medical inspection as a substituting strategy of control meant that irrespective of disease at the original port, if passengers, vessel and goods were assessed as healthy they were allowed to disembark. 'The system', it was noted 'isolates the sick, allows the healthy passengers to go free after taking their names and destinations, and detains the ship only long enough to permit of the necessity of disinfection'.[10] In most countries, then, quarantine was in decline compared to other preventive measures.[11] As Dr Charles Mackellar put it to the Royal Society of New South Wales in 1883, 'a much modified system seemed to be viewed with most favour – rather a system of inspection and purification than one of detention during the incubatory period'.[12] Authorities for England and Wales had abandoned strict maritime quarantine at their ports with respect to cholera in 1873 (but not for yellow fever or plague) and had put in place a system of medical inspection. From 1896 in England and Wales, medical inspection became the practice with respect to all infectious diseases.[13]

In his book *The Invention of the Passport: Surveillance, Citizenship and the State*, John Torpey studies the development of passport controls 'as a way of illuminating the institutionalisation of the idea of the "nation-state" '.[14] He begins his fascinating book with the role of the passport in regulating movement in *ancien régime* and then revolutionary France, that is, at the birth of the nation-state. Documents regulating movement had been required earlier, he shows, but in this summary of premodern as well as his detailed work on the modern's state's monopolisation of the authorisation of movement, inclusion and exclusions, quarantine and infectious disease measures are barely mentioned.[15] Yet as much of my book shows, infectious disease, migration and movement have long been jointly regulated by mercantilist states and by modern biopolitical states. More to the point, this regulation produced identity documents showing clean health status of individuals seeking free movement. These are in fact old devices, suggesting the combined histories of commerce and quarantine, and one appropriated by new nation-states alongside, and increasingly as part of documents of citizenship and the passport. As we have seen, for example, documents showing vaccination and even inoculation history and status have been intermittently both produced and required by states, as individuals seek entry or exit. Over the twentieth century, the 'bills of health' once granted according to the health of a vessel, were increasingly individualised and became bills of health for each family, and for each person, classifying and subjectifying not only in terms of nationality and race but in terms of cleanliness, fitness and health. Often, as I discuss in the next chapter these were conflated categories as the biopolitical state became increasingly racialised, and as an 'international hygiene' became increasingly eugenic.

Quarantine, both as procedures for the entry of ships, their goods and people, and as we have seen as an emergency measure in times of epidemic for domestic populations, was well established in each of the Australian colonies. The original New South Wales Quarantine Act (1832) had been passed as a response to cholera in England, when the Imperial government was still sending transported convicts as well as settlers. Over the 1880s and 90s and as part of the discussion about new Commonwealth responsibilities after 1901, many authorities agreed that Australia should both harden and standardise its maritime measures, that is, eschew the reduction of quarantine measures which characterised 'our Imperial sanitary authorities'.[16] Over these decades, Australian authorities devised and passed new quarantine acts, established complex new quarantine administration and practices, sites, regulations and enforcement agencies. The system of detaining *all* passengers if infectious disease had been present *en voyage* in *any* of them was generally advocated. By the same token, it was unsustainable in Australia to administer quarantine strictly on the principle of health of the port of embarkation, simply because many disease endemic in, say,

Ceylon, were not only not endemic on the Australian continent, but entirely absent. This would mean, logically but impossibly, that every ship would be quarantined.

In an under-recognised way, discussion about co-operation between the Pacific and Australian colonies (Fiji, New South Wales, Victoria, South Australia, West Australia, Queensland, Tasmania, New Zealand) on the question of quarantine became entwined with the movement for their political federation. Ideas had been mooted for the joining of the Australasian colonies (at various times including New Zealand and Fiji) from the middle of the century, initially by Colonial Secretary Earl Grey in 1847. In 1885 the Imperial Parliament passed the Federal Council of Australasia Act creating a formal body for the colonies to confer and possibly to legislate jointly. Various councils, conventions and referenda were held in the 1880s and, '90s, and in 1900 the new constitution of the Commonwealth of Australia was passed by Act of Imperial Parliament, and assented to by Queen Victoria. In 1901 the new nation was constituted, joining the colonies of New South Wales, Victoria, Tasmania, South Australia, West Australia, and Queensland.[17]

The Federation Conferences of the 1880s and '90s are famous in Australian political history, but there was a less recognised yet contemporaneous string of intercolonial Australasian conferences dealing with sanitary questions, in particular with the desirability of co-operative federal quarantine.[18] Not only military defence (which is usually referred to in the historiography) but the related domain of maritime quarantine was one of the early rationales for administrative cooperation between the colonies. In 1883, the president of the New South Wales Board of Health Charles Mackellar published an oft-cited article 'Federal Quarantine', in which he presented an argument for a federal system, based on many years experience of inconsistent and inadequate quarantine measures by different colonies. In his opinion, agreement on quarantining measures and some level of joint administration was imperative 'for our common weal', as he put it.[19] Mackellar's efforts centrally informed the important 1884 Australasian Sanitary Conference held in Sydney, at which it was resolved to pressure the colonial governments to consider a jointly administered quarantine system accompanied by uniform as well as compulsory vaccination across the colonies. The 1884 Conference resolved that a Quarantine Bill be drawn up, and be recommended for adoption by each of the Colonies under the title of The Federal Quarantine Act of Australasia.[20]

This did not occur under the colonial system of rule, but the power of quarantine transpired to be the one and only constitutionally named public health power granted to the new Commonwealth (section 8). All other health powers were left with the States, largely to be implemented in concert with local government.[21] In 1908 the *Quarantine Act* was passed, and this brought about the creation of the Federal Quarantine Service in 1909, a bureaucracy significantly located in the Department of Trade and Customs. Already, then, there was a telling tracing of concerns of health and population with economy and security. The Commonwealth Department of Health was established in 1921 and originated institutionally from this Federal Quarantine Service.[22] Both in terms of the Constitution, then, and in terms of this bureaucratic genealogy, the administration of the public health of the nation had its origin in quarantine.

From the original quarantine power many other responsibilities were argued for and justified. For example, the effective co-ordination and assessment of quarantine measures were immediately taken to require (or made to require) the collation of data: vital statistics. At the 1904 Commonwealth Quarantine Conference, where the specificities of a Commonwealth approach were nutted out, it was recommended that the work already carried on by the State statisticians in the collection of vital statistics should be made available to the Director-General of Quarantine, who could collate them for the nation.[23] There was a marked expansion of the reach of Commonwealth health powers in the hands of an extraordinary public servant, Dr John Howard Lidgett Cumpston, Director-General of the Quarantine Service from 1913, and long-time Director-General of Health. On appointment he controversially enlarged these quarantine powers *vis-à-vis* the States, and expanded the scope of responsibility for health that might legitimately fall to the Commonwealth within the confines of the Constitution. Cumpston wrote that the immediate reasons for the creation of a Commonwealth Department out of the Quarantine Service were the influenza pandemic in 1918–19, health issues related to the intake of refugees and the return of soldiers after the First World War, as well as the consideration of the 1916 Report on Death and Invalidity by the Commonwealth.[24] By 1921 when he successfully argued for departmental status, the functions of the Department of Health were ordered as follows: Administration of the Quarantine Act; the investigation of causes of disease and death; the establishment and control of laboratories for this purpose; the Control of the Commonwealth Serum Laboratories and the commercial distribution of the products manufactured in those Laboratories; the collection of sanitary data, and the investigation of all factors affecting health in industries; the education of the public in matters of public health; the administration of any subsidy made by the Commonwealth with the object of assisting any effort made by any State Government or public authority directed towards the eradication, prevention, or control of any disease; the conducting of campaigns of prevention of disease in which more than one State is interested; the administrative control of the Australian Institute of Tropical Medicine.[25] By 1928, the Commonwealth Department of Health managed a medical officer at the High Commission in London and Chief Quarantine Officers in each State. There were separate Divisions of Marine Hygiene, Industrial Hygiene, Public Health Engineering, Epidemiology, Veterinary Hygiene, and Plant Quarantine, a Cancer Division and a Division of Tuberculosis and Venereal Disease. The Division of Laboratories managed the Commonwealth Serum Laboratory in Melbourne, and eight other Public Health Laboratories.[26]

Cumpston lived between 1880 and 1954, thus working, writing and thinking precisely within the decades which were most crucial in the development of bureaucracies of public health, not only in Australia, but in North America, France, and Britain. This was, as we have seen, the period of new social government in the forms of new liberalism, progressivism, or welfare. Cumpston's extraordinary career also traversed the formative moment of Australian nationalism, the consolidation of a new federal Australian political structure, and the development of independent Australian relations with other sovereign nations, as well as international bodies like the League of Nations Health Committee and the Rockefeller Foundation.[27] He was educated in and worked through the period which saw bacteriology and laboratory-based medical and public health knowledge more or less displace older theories and methods, and he saw himself very much as a 'modern' and progressive thinker in questions of health.[28] Educated in medicine at the University of Melbourne, he travelled to Japan and the Philippines where he was introduced directly to tropical medicine scholars and issues. He went to London in 1906 to undertake a diploma of public health, visiting the Lister Institute of Preventive Medicine and the Pasteur Institute, and attending various international conferences including the Berlin International Congress on Hygiene and Demography in 1907. In 1908 he took a position in the Western Australian public health service, later moving to the position of Director of the Quarantine Service where, in the rigid segregation of Sydney in a smallpox outbreak in 1913, he immediately flexed new Commonwealth muscle over the States, putting their relative powers and jurisdiction to the test. Not only was Cumpston singularly influential in the shaping of public health and national schemes of hygiene over the first half of the century, he wrote a stunning series of medical histories, compiling vast quantities of data into the epidemiology of almost every disease of import at the time, as well as comprehensive histories of his own Department.[29] Cumpston, like many of his public health colleagues, needs to be seen understood as a 'progressivist', as Michael Roe has argued, imbued with the values and mission of early twentieth century national efficiency.[30] In Cumpston's hands, as well as other experts driven by the connections between health, colonisation, nation and government (for example J.S.C. Elkington, Raphael Cilento, Cecil Cook) health and hygiene were readily linked to nationalism and national efficiency. And as I discuss in the next chapter, 'hygiene' in this period moved fairly seamlessly into 'racial hygiene' and the foundational national policy of white Australia.

The island-nation: marine hygiene and the national border

Quarantine, more than any other government technology *is* the drawing and policing of boundaries. Quarantine and nationalism imply each other because both are about the creation of spaces. They determine an internal and an external, often nominated as clean and dirty, through the administration of a boundary. Boundaries are required for the creation of nations in a modern Western sense, and quarantine is in essence the putting of these boundaries to a particular use by the administrative nation-state. Not infrequently, quarantine and national administration produce and monitor the same space: that is, the border of a nation has often been where a quarantine line was drawn. This same border might well have a military, political and economic significance; the place of potential invasion and defence as well as commercial traffic and exchange. Sometimes such a border recommended itself for all of these functions – nation-marking, quarantine, commercial exchange, defence, as well as immigration regulation – because of a particular geography: a range of mountains, a river, a sea. Geography in part determined the possibility and efficacy of quarantine measures, this in itself underscoring the arbitrary nature of national lines. Often enough, permanent quarantine lines were conflated with national borders because, simply, those borders were already there. National borders were places where, by definition, government powers were already exercised; the border was already policed and administered; it already had a governmental and bureaucratic significance. The microprocesses of government which monitored the exchange of goods, the inspection of people for disease, the regulation of prospective immigrants, intensified the effect and significance of already established borders.

But national borders, like quarantine lines, are not necessarily places of total exclusion or enclosure. 'The frontier is both an opening and a closing', wrote Edgar Morin. 'All frontiers, including the membrane of living beings, including the frontier of nations, are, at the same time as they are barriers, places of communication and exchange. They are the place of dissociation and association, of separation and articulation.'[31] Many borders aim to regulate and control movement, flow and exchange, not stop it all together. This is a useful idea for quarantine boundaries in particular, as, unlike military defence for instance, quarantine is a technology of government which is as much about regulating *entry* as it is about keeping people and microbes out. National quarantine lines are often less an impassable barrier than a net which screens.[32] Indeed the metaphor of the net, not uncommon in early twentieth century conceptualisations of maritime quarantine, captures the idea of 'screening out' unwanted people and things, while permitting entry to the fit and healthy. It also captures the constant anxieties about 'holes' and 'leakiness', about the undesirables slipping through.

The idea of quarantine was one of the means by which nations imagined their integrity; quarantine lines made otherwise often abstract national or colonial boundaries very real. This has been richly argued by Alexandra Minna Stern of the US-Mexico border, and by Heather Bell of the colonial constitution of the Anglo-Egyptian Sudan.[33] In these cases, the boundary was a hyper-administered, but essentially arbitrary line on contiguous ground. This facilitated legal and illegal crossings and re-crossing of customs, immigration and quarantine borders. These borders were, to cite one political geographer, the places where 'the vertical interfaces between state sovereignties intersect the surface of the earth'.[34] Quarantine in Australia was similarly important to the creation of a sense of national boundary, but in a markedly different way. As an island-nation the 'vertical interface' was never a direct interface with another sovereign entity. The Australian national border did not intersect with earth but with water: it was never a line on the ground, but a vast expanse of ocean to the west, east and south, and, importantly, a not-so-vast expanse to the Asian north. Writing of the international sanitary conferences, W.F. Bynam has noted, 'The AIDS pandemic has reminded us starkly that no man or his nation is an island'.[35] Bynam means, of course, that microbes do not respect borders, that we live in a world where borders are crossed and recrossed with increasing rapidity, a world, which is becoming increasingly borderless with some significance for the movement of communicable disease. Yet his comment gets to the heart of the matter for the new Australia. The nation *was* an island, and an isolated one at that. A national maritime quarantine line was created by the Commonwealth government and its new Quarantine Service: this outlined the new island nation, enclosed it, segregated it, and integrated it.

The geographical borders of the continent seem now to recommend themselves in an obvious and commonsense way for the integration of the previously separate British colonies. Yet as historian Helen Irving has shown, this was by no means the only or the obvious territorial configuration of the Australasian colonies. In some incarnations of a federal plan, for example, the crown colony of Fiji and New Zealand were incorporated, in others Queensland and Western Australia were excluded: 'the nation's borders were not predetermined. They too had first to be imagined'.[36] The maritime quarantine line was one important way of imagining Australia as a whole, as the island-nation it was. A string of new federal quarantine stations dotted the thresholds of the nation and a 'net' strung between them was thought of as a protective barrier – a defence at the frontier. This metaphoric 'net' was represented on maps, and as a legal line, but it was also concretised in the day-to-day practices of customs and quarantine (and, as we shall see, immigration) officers and agents. The geography which formed the nation – the island – also protected it from diseased others.

> Civilised countries nowadays keep themselves free of dangerous epidemic diseases by keeping them out. Quarantine has been organised, as all public services have been organised, until it is now a very fine-meshed net stretched round a country so that all disease, whose introduction might have serious consequences is caught and stopped from entering.[37]

Countless maps of Australia were produced by Cumpston's bureaucracy which mapped, in epidemiological style, the spatial clustering of infectious disease. The clustering of disease along the coastline-margin at once outlined Australia and underlined the necessity for quarantine.[38] In this process, attention was drawn to the vulnerable points of entry and exit. New quarantine stations were deemed imperative for Thursday Island and Townsville in the north, Albany for ships coming from Europe and India in the west. The major cities had long functioned as the quarantine focus, in a geographic imagining structured by interests and imperatives of the separate colonies. But once the island of Australia began to imagined as a commercial and political whole, the peripheral sites of Thursday Island and Albany became newly significant. That is, if Sydney or Melbourne or Perth were understood as 'points of entry' in a colonial imagining, Thursday Island and Albany were the points of entry in a federal imagining, points of vulnerability and therefore points to be secured.[39] In particular, the tiny Thursday Island, the northernmost part of Australian territory bordering on what is now Papua New Guinea and East Timor, held a new national significance as a truly threshold site.

The quarantine 'net' protected and defended, and thus in part created, an Australian island-nation imagined as uninfected and 'virgin'; that is, as pure and white. There was considerable investment in medical, epidemiological and governmental discourse in imagining the island-nation of Australia as 'new', 'clean' and 'healthy'; as the only continent free from endemic forms of certain infectious diseases. In discussions on federal quarantine in 1895 for example, it was argued that a 'code of sanitary regulations for Australia ... would give the colonies the best chance of still retaining the proud title of the virgin continent'.[40] In arguing for the maintenance of strict quarantine, and not following the British and European path of diluting, or abandoning maritime quarantine, this image of the continent as free of disease, as 'clean space' was mobilised over and over again. The 1884 Sanitary Conference reported, for example: 'we have here a virgin country that has never been visited by these disorders, and the interests involved are so vast that I do not think we can be too rigorous in dealing with cholera'.[41] The clean and white space of Australia is graphically represented in a 1912 Commonwealth Report on Quarantine, in which several maps indicated the global distribution of cholera, smallpox, and plague marked in black. Australia, at the centre of the map is free of disease, white and clean (see Figure 5.1).

Australia was imagined as having, indeed did have, a geographic protection by virtue of its global distance from Old World centres of disease: 'In Australia we are happily situated at such a distance from other countries that most diseases have time to develop *en voyage*. Nature has thus established a sort of prophylactic quarantine'.[42] While this distance from the Old World was in this context seen to be fortunate – Australia was new, pure, healthy and white – it was precisely this same distance which located white Australia in an Asia constructed as dirty, diseased and all that was not white.[43] In an article titled 'The Protection of our Frontiers from Invasion by Disease', Cumpston

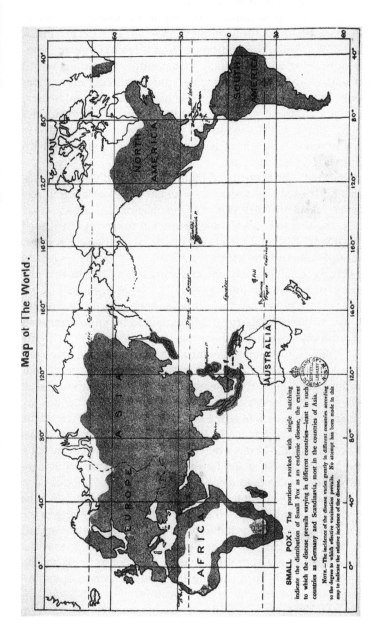

Figure 5.1 Australia – clean and white
Source: W. Perrin Norris, *Report on Quarantine in Other Countries and on the Quarantine Requirements of Australia*, Government Printer, Melbourne, 1912. Courtesy, Mitchell Library, State Library of New South Wales

warned: '[Australia] is fortunate in that it is protected on all sides by the open ocean, and unfortunate because of its close proximity to numerous endemic focci of communicable disease: Egypt, India, China, Malay Peninsula, the Philippine Islands are in close and constant communication with Australia.'[44] Quarantine in Australia was in large part shaped by this particular geographics. While in many cases the presence of endemic disease in these places was epidemiological and microbiological fact, it was never just that to Australian authorities and experts. Rather, as I discuss in the next chapter, this was always (and arguably still is) a racialised geographics heavily connected with the imperative of 'white Australia'.

The complicated logic of purity and impurity, susceptibility and resistance which I discussed with respect to vaccination and isolation in previous chapters, came into play as part of the imagining of the geo-body of Australia. The very 'purity' of the continent in some ways weakened it, rendered it vulnerable and susceptible to infectious disease. For some, this made large scale and compulsory vaccination absolutely necessary. Precisely because smallpox was not endemic, it was argued that the population of the country, Indigenous and non-indigenous had no natural resistance to it. Further, if vaccination was compulsory, quarantining ships and immigrants for smallpox would be unnecessary, an argument strongly favoured by commercial interests. The highly respected J. Ashburton Thompson argued, with reference to a disastrous epidemic in Montreal, that quarantining for smallpox was simply outdated and irrelevant when the 'true protection' of vaccination was at hand.[45] At the 1884 Australasian Sanitary Conference, attended by the Australian colonies as well as the government medical officer of the Crown Colony of Fiji, it was agreed that 'by producing personal insusceptibility of the people at large by vaccination ... the necessity for a Quarantine against small-pox and for the detention of persons not actually sick would be removed. Indeed Quarantine against small-pox is, at this date, an anachronism; yet it is absolutely necessary that it should be maintained with great strictness in all part of Australasia as long as the population in any part remains imperfectly vaccinated'.[46] Others argued to the contrary and along the lines discussed in Chapter 1, that the vaccine itself was a contagion, that the population of Australia should not be vaccinated at all, thus keeping the continent and population pure and isolated, and that the strictest measures of maritime quarantine should be relied upon.[47] Indeed it seems that amongst anti-vaccinationists internationally, Australia had a misconceived reputation as an unvaccinated country that also had no smallpox. In the 1920s

Cumpston's new Department received numerous requests especially from US public health bodies as well as anti-vaccination organisations, to supply information about the legal situation and the implementation of vaccination as well as quarantining measures in Australia.[48] Altogether, it is important to recognise the extent to which the relative absence of smallpox on the Australian continent meant that vaccination and quarantine were often discussed together, as twin issues, involving complicated and multidirectional lines of hygiene. For Cumpston, indeed, vaccination and quarantine were not alternatives to one another, but rather a first and second line of defence.[49]

This military language of defence and invasion was a common way of comprehending and articulating public health work in the early twentieth century.[50] In the new Australia, this contributed to the intense nationalism of the period. While nothing quite mobilised nationalist sentiment like war, invasion by disease not uncommonly stood in for the threat of actual invasion. Quarantine, then, was one strategy by which the new government could imagine itself at war with its region, as it were, thus defining the new Australia geographically, politically and racially against Asia. Although smallpox and plague were sometimes understood to be 'imported' (the secondary discourse here was commercial), most commonly they 'invaded'. Cumpston wrote that 'hygienists in Australia will look on the seaward frontiers as the places which must be fully manned and equipped with the most modern armamentaria in order that the possibility of invasion by disease shall be reduced to an absolute minimum'.[51] The cultural geography of the Quarantine Station at North Head in Sydney is telling here. Always a place where the administration of immigration and the production of national health came together, it was also a place designated to the military defence of the nation: the peninsular came to be shared with the Commonwealth Army. Military forts for the domestic protection of the continent in war and the old Quarantine buildings still sit side-by-side in the liminal space between the city and the ocean. They are concretised examples of this mutuality of military and biomedical discourse, and of the ever-present significance of Australia's island-ness, the oceans as both isolation and security, at once permitting and requiring defence.

Imagining Australia in space and time

While the original meaning of quarantine implied the primary significance of time – forty days isolation – in the late modern period the prominence of spatiality in problematising disease produced a notion of quarantine which always turned on questions of both space and time, and, if anything attributed a primary significance to geography, the element of 'isolation'. Epidemiology as an evolving modern knowledge was also driven by the confluence of space and time. On the one hand, it drew from the techniques of urban spatial mapping and classifying which characterised the nineteenth century social surveys from Chadwick's report, to Snow's maps of urban cholera distribution, to the Booth map of London class.[52] On the other hand, epidemiological knowledge was driven by the axis of time, which in part was used historically and in part predictively, to assess risk in the future. Epidemiological knowledge, especially in the nineteenth century was organised characteristically through the all-important 't' axis, the interpretive usefulness of incidence-over-time, rather than prevalence at a cross section of time.[53] This confluence of time and place is neatly suggested in the title of the nineteenth-century epidemiological journal *Geographical and Historical Pathology*. The government technology of quarantine and the expert discipline of epidemiology were epistemologically similar, both working from and toward knowledge of individual, population and microbe in terms of geography and temporality: the natural history of a disease in a person as well as the location and movement of that individual; the natural history of an epidemic in a population as well as its mapping; and the all-important question of incubation period and the tracing of contacts within that period of time.

Because of the insular geography of the new Australia, the continent-nation was strongly imagined by experts in terms of this space-time delineation characteristic of both epidemiology and quarantine. Put another way, epidemiology and quarantine were joint knowledge-practices through which 'Australia' was imagined. The rationale for the 1884 Sanitary Conference was proclaimed thus: 'The countries which together constitute Australasia are separated from the rest of the world by a barrier of time-distance'.[54] The time-distance between various European nations, by contrast, was so minimal that 'so far as quarantine was concerned, [Europe was] merely one country'.[55] The pressing question for public health bureaucrats and the administrators of quarantine, was, of course, the number of days between various ports and the ports of arrival in Australia. This time was placed against, and drew its significance from, the time in which any given disease revealed itself symptomatically, as rash, fever, or pustules: its incubation period. For the so-called Far East, however, this was not necessarily the case, and health, customs, immigration and quarantine officials were constantly concerned about ships arriving in Thursday Island, Queensland ports and in Darwin, from the nearby Straits Settlements, Singapore, China, and the Dutch East Indies.

The mapping of Australia for quarantine purposes drew much attention to this temporal dimension. The isochronic charts developed by the Commonwealth Department of Health located Australia explicitly in both space and time relative to its region, marking out the time-distance by ship between the northernmost part of Australia and these Asian ports further north (see Figure 5.2). The isochronic chart once again reveals the special significance of Thursday Island in the new geo-body of Australia, as the liminal point of entry and exit, of exchange and articulation with Asia. On the other hand, ships from Europe came via India and Ceylon and for much of the nineteenth century, when ships were slower, there was a certain confidence in the natural protection of the 'barrier of time-distance': 'Australia is in a fortunate position. Our principal ports have between them and the big Old World centres of population and disease a sufficient sea space to enable all disease to make itself known before arrival here'.[56] What had been seen to protect Australia, to give it its purity (but also therefore its vulnerability) was the fact that the time to sail from European, African and Indian ports from the west and American ports from the east, to Australia was more than the time taken for smallpox, cholera or yellow fever to appear in an individual. But this was changing. Australian public health experts kept a very close microbiological eye on developing understandings and estimates of incubation periods of cholera, yellow fever, smallpox, plague. And as I discuss below, they kept an extremely close epidemiological eye on communicable disease at Indian ports and at Colombo. In particular, health bureaucrats like Cumpston monitored with concern the increasing speed with which steamers were arriving at Australian ports.

The technology of quarantine in the early twentieth century was shaped by the central questions and developments of modernity: the increasing speed and changing modes of travel, the complexity of communication, the shrinking of space and time, the development of global information paths enabled by wireless and by telegraph.[57] Every year the protective area around Australia was not only shrinking, but changing form: 'As the speed of ships advances, the risks to which Australia is exposed must increase ... modern travel improvements mean danger to Australia'.[58] And the logic was that if the 'natural protection' of the time-distance barrier was being breached or weakened

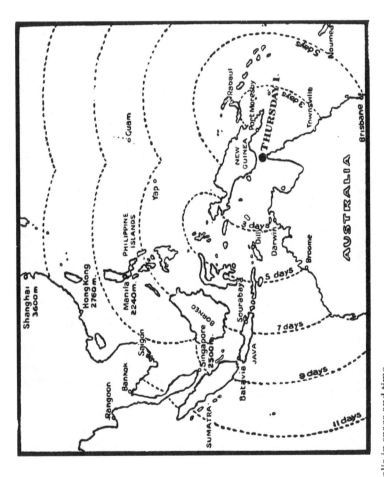

Figure 5.2 Australia in space and time
Source: Isochronic Chart, from *Health*, 5 (1927): 46. Courtesy Mitchell Library, State Library of New South Wales

technologically, then the government defense of quarantine needed to be strengthened administratively, legally, and imaginatively. By the 1920s air travel changed this temporal and therefore spatial understanding of microbial danger again. The 1919 flight of Ross Smith From England to Darwin prompted a change to the meaning of 'vessel' in the Quarantine Act.[59] And a 1928 issue of *Health*, the Commonwealth Department's intriguing internal journal, detailed the significance of Kingsford Smith's flight from San Francisco to Brisbane. It had 'perhaps done more than previous overseas flights or even wireless to break down our ideas of time and space. Our sense of security from a geographical isolation has received a shock'.[60] Air travel raised questions of compulsory vaccination once more, and precisely in terms of the 'removal of the old barriers of time and distance between Australia and other countries', which resulted from air travel.[61]

The cultural historian Stephen Kern notes the way in which air travel in the early twentieth century was written about as dissolving national borders. Air travel 'sundered protective frontiers and created new spatial dynamics'.[62] Aeroplanes reordered the significance of islands as places of immunity or isolation. Gillian Beer writing of Gertrude Stein's musings on air travel, suggests that this reordering 'does away with centrality and very largely with borders. It is an ordering at the opposite extreme from that of the island, in which centrality is emphasized and the enclosure of land within surrounding shores is the controlling meaning'.[63] All the more reason, then, to shore up the borders, to reinforce them and make them visible and meaningful with the multitude of micro-practices of quarantine enforcement, in the Australian case.

But the modern technology which brought dangerously invisible microbes to Australia with increasing rapidity also provided potential new means for resistance and control in the space-time race. In particular, revolutions in communication and information collation and generation were brought to the battle against communicable disease. Receiving, distributing and generating information about global and Australian cases and epidemics of communicable disease became a primary function of the Commonwealth Department of Health under its quarantine power. The Department became part of a new network of health organisations which aimed to make epidemic disease globally traceable, predictable and therefore, it was imagined, preventible.

These international organisations had their origins in the international sanitary conferences of the nineteenth century. The Office international d'hygiene publique was established in 1907 with conventions and procedures aimed against the spread of cholera between European nations and between Europe and America. The Rockefeller Foundation was established in 1913 to promote public sanitation as well as ideas about scientific medicine. With the new possibilities of, and drive toward internationalism after the First World War, the range of agreements, organisations and communications proliferated. Various international incarnations of Red Cross Societies were reinvigorated. Most importantly after 1921 the Health Committee of the League of Nations was created. With the co-operation of ratifying governments, the latter had the resources and the rationale to create a 'Service of Epidemiological Intelligence and Public Health Statistics' in Geneva, creating an international epidemiological database which the World Health Organization inherited after the Second World War.[64] Particularly significant in this respect was the establishment by the Health Committee of its Eastern Epidemiological Bureau in Singapore (1925) funded in part by the Rockefeller Foundation.[65] In the next chapter I discuss all this as 'International Hygiene' and relate it to the growing network of immigration restriction acts. For the moment, though, I want to suggest Cumpston's Department as deeply enmeshed in this network, and the knowledge produced not only justified his ever-growing Department, but was part of an imagining of quarantined Australia in space and time.

Between participating nations and colonies passed a vast amount of what was called 'epidemiological intelligence'. The information which indicated that a ship or an individual was infected, needed to reach quarantine bureaucrats and administrators *before* the infected vessel or person reached the borders of the country. Thus any given health bureaucracy would have knowledge of potential and actual cases of communicable disease prior to a ship arriving in their port; the need for quarantine was known in advance. Cumpston detailed the process thus:

> Each week a code message is broadcasted by radio from Saigon and other wireless stations. This message contains the substance of information received regularly over a very large area extending from Greece, Egypt, and the East Coast of Africa, through Arabia, Persia, India, Ceylon, Siam, China, Japan, Siberia and the East Indies to Australian, New Zealand, Fiji and the Western Pacific Island groups ... The message is received each Friday in Sydney, decoded in the Sydney office of this Department and then relayed to the Central Office at Canberra and to all branches of the Department.[66]

He then produced an example of the message in code and decoded.[67] The question of the speed of this intelligence, of this international communication, was crucial to the public health defenders of the nation such as Cumpston. For it was imperative that the speed of the information be greater than the speed of the disease (that is, of the ships). Information literally jumped over (via airwaves) or under (via cable) the ships which carried the infected case, the carrier or the contact. This information increasingly made redundant the old 'Bills of Health' which each vessel was granted on entry to a port. Indeed these were abolished by the Australian government in 1929.[68] This notion of 'intelligence', of coded communication between governments and their agents betrays the politico-military genealogy of the Bureau, the Health Committee and the League itself. As Manderson has written: 'whilst the primary enemy in epidemiological surveillance was disease – especially plague, smallpox and cholera – the implicit enemy was the outsider, and tensions of autonomy, empire and territory were played out literally and metaphorically in the work of the Bureau'.[69] 'Intelligence' intensified the shared meanings between the defence of the public health and the politico-military defence of the nation; the significance of 'health' for 'security', and the significance of 'security' for 'identity'.

Conclusion

Quarantine was a significant aspect of the geo-body of the new twentieth century nation, Australia. But it was also necessarily about the multiple crossings of the nation's new boundaries, that is it produced *inter*national relations too. Nationalism, always involves a 'two-way identification'.[70] Quarantine measures were an important site where concepts and practices of international relations and border control developed over the nineteenth century, as nation-states were created and became more regulated and policed. As I discuss in the next chapter, quarantine measures often doubled with diplomatic, immigration and customs measures, all of these being means by which nations defined themselves and imagined themselves as (more) secure. *National* administration of quarantine and disease prevention forced the evolution of *inter*national relations in increasing administrative complexity and sophistication. As the Australasian Sanitary Conference foreshadowed in 1884: 'the health of any nation is dependent not only upon its own efforts, but that the obligation to carry on the work of sanitation is in the nature of things an international obligation'.[71]

As border sites, national quarantine lines – not least as lines of immigration – are always significant places, dense administrative sites which often betray those 'differences' which are permissible within a nation at any given moment, and those which are deemed undesirable. All this, of course, creates identity. But the Australian case is unique and especially telling here because this was a nation being created and consciously creating itself – constitutionally, physically, administratively, spatially, and as we shall see, racially – at the same time as health measures were newly implemented at a national level. Maritime quarantine was not only part of this process, it enabled an imagining of the new nation as the island it was, as an integral whole. The geographical borders of the island, which bounded the new nation and made it one, *were* the borders pressed into use in quarantine. All of this administrative focus, this mapping and enforcing of the quarantine line which literally outlined the nation, worked toward the production of Australia as a 'geo-body'. Maritime quarantine was part of a nationalism figured not only through culture – cultural difference from Asian countries and in a different way, from Britain – but a nationalism figured through an insular territoriality.

6
Foreign Bodies: Immigration, International Hygiene and White Australia

In this chapter, I look again at the borders of the geo-body of the new nation of Australia and examine how the quarantine line was also a racialised immigration restriction line. The maritime quarantine line secured the (insecure) geo-body and civic body of white Australia as part of the racialised defence response to an 'invasion narrative' which governed much law, literature, culture, and policy of the early twentieth century.[1] This was the moment when the health of the Australian nation – national purity – was being realised as racialised aspiration. It was a moment when a racial politics was institutionalised and legitimated, that is, rendered into law: the Immigration Restriction Act and the Pacific Island Labourers Act, both passed in the first year of national government, 1901, formed the basis of what became known popularly as the white Australia policy. Here, I draw the relations between white Australia and international hygiene

The deep cultural as well as legal connections between quarantine measures and immigration restriction measures formed part of an 'international hygiene', as it came to be called in the interwar period, interlocking legislative tools for the inspection and restriction of imperial and global movement, mainly of non-British, non-white people. Beginning around the 1880s, these restrictions tightened until the post Second World War revisions of sovereignty, race discrimination and human rights covenants. These developments may be understood as 'international biopolitics': the modern nation implementing not only an 'administration of life' internal to itself, as biopolitics is conventionally considered, but an administration of life oriented externally: the regulation of immigrants, refugees, indentured labourers, tourists, students.[2] International biopolitics is that which governs and administers 'life' and 'population' across and between sovereign states, not just within them. One line of the genealogy of this international biopolitics can be traced to the sanitary conferences of the nineteenth century, the race- and health-based immigration restriction of the 1880s to the Second World War, and the gathering of these systems under the rubric of international hygiene.

The Australian instance is important in this international history of borders, hygiene and national/racial identities because it was essentially the test-case on race-discrimination and national sovereignty at the Peace Conference after the First World War. The 'white Australia policy' therefore became, and remains, internationally infamous. Yet it is important to contextualise Australian immigration law and policy within this much larger international hygiene, this developing international biopolitics. The Australian case *is* unique, but not, I argue, because 'coloured aliens' were excluded from its territory and body politic: such legislative moves of race- and health-based nation building were more ordinary than extraordinary for the period. Rather its uniqueness stems from the simultaneity of the adoption of such racial exclusions *with the moment of* nation-formation. The racial exclusions were the defining act of the new nation in 1901. Thus, although most other colonies of British (white) 'settlement', as well as nations in North and South America also legislated to exclude racially, that this happened at the moment of nation-formation in the Australian instance makes the connections between racial hygiene, national hygiene and the constitution of a white civic body by exclusion (as well as selective inclusion) both formative and tight.

To draw connections between race, nation and hygiene is in many ways an unsurprising if necessary set of historical observations. This is all problematically familiar in the histories and historiographies of numerous white 'settler' societies and the Chinese diaspora, as well as the national and racial hygiene programs of Nazi Germany.[3] Here my purpose is not simply to offer a further rendition of the way in which metaphors as well as policies of (racial) purity and contamination played out in another context, but also to argue for the technical and legal ties between health management and race management, especially medico-legal border control. The question of 'foreign bodies' in Australia demonstrates the modes of spatial governance which racial discourse and health discourse shared, especially in eugenic renditions, a topic I take up in both this chapter and the next. The state apparatus for maintaining the borders of the nation, for managing quarantine and for implementing immigration restriction were more or less the same. The example of white Australia and its production through combined quarantine/immigration measures shows how 'whiteness' was technically implemented as well as culturally imagined through, and in concert with, public health. 'White' was not only a racial identification in this period, it also signified purity, hygiene, cleanliness.[4] 'Whiteness' then, was the business of public health, and immigration regulation was one site where this national purity was sought and implemented. By these means the Australian population was shaped, borders were ostensibly secured and the health of the nation – national purity – was pursued by government as a racialised objective.

When Dr A. Wallace Weihen spoke to the Australasian Medical Congress in 1911 he framed his talk, 'The Medical Inspection of Immigrants to Australia', in terms of a segregative imperative longstanding in public health but given new expression through eugenic ideas: 'In these days of eugenics we must recognise that, apart from education, any attempts to improve race-stocks are limited to two main directions: First – Segregation within the limits of a country. Second – Segregation by refusing entrance to undesirables from without'.[5] Throughout the book, I have examined several sites of 'internal segregation' in operation from the 1880s. Here I focus on, as Weihen put it, the segregation of 'undesirables from without'. This worked simultaneously through joint biopolitical measures of quarantine and immigration restriction, first, by excluding many 'coloured' people through racial pathologisation, and second, through border screening of healthy from unhealthy Britons. The classificatory and segregative impulse – the clean from the dirty, the fit from the unfit, the desirable from the undesirable – crossed over the domains of race management, public health and eugenics, but in no straightforward way. The Australian exclusionary immigration acts of the turn of the century were intended for, and indeed effected the exclusion of 'coloured aliens', especially Chinese people, as foreign to the territory and body politic of Australia: this was the white Australia policy. Thereafter, however, the whiteness of white Australia came to be as much a eugenically defined as a racially discriminatory pursuit, seeking in a multitude of sites, the enhancement of the quality of the whiteness of the nation. This was the racialisation of the biopolitical state.[6] The 'hygiene' at issue here was the border-exclusion of the biologically, mentally or morally unhygienic and contaminating Briton: these were the 'undesirables' to whom Weihen referred. Unhealthy Britons, not just racially other 'coloured aliens', were also 'foreign bodies', imagined through, but not identically situated in, the same broad cultural and legal system of hygiene, race and nation.

In the northern zones of the nation, however, even the fittest and healthiest white-Briton was marked by the powerful discourse of tropical medicine and hygiene as climatically, geographically and constitutionally out of place. This is the final, ironic turn in thinking about Australian health, race and immigration politics in terms of 'foreign bodies'. Tropical medicine, as well as the growing literature on 'international hygiene' of the interwar years, studied and problematised global racial distribution and redistribution through migration and systems of indentured labour. The un/desirability of such movements were partly governed by economic needs, and partly by strongly held ideas on the necessary relation between racial constitutions and environment or climate: there were places where white people belonged biologically-speaking, and places where they did not. This had long posed a problem for British imperial aspirations, a problem for the 'diaspora' of white British bodies which constituted the British Empire. As one commentator put it in an article titled 'The Influence of Race on Climate', the English are 'masters, for the present at any rate, of countries situated in almost every degree of latitude'.[7] This came to be dealt with through the institutionalised imperial discipline of tropical medicine. In one of the great ironies of the modern colonial period, the exclusion of 'coloured' labour from the colony-nation of Australia in the obsessive pursuit of whiteness, came up against a counter-problem: the idea entrenched in the colonial discourse of tropical medicine that 'white man' is foreign to the place of the tropics. After the exclusionary immigration Acts in Australia, their vehement defence internationally after the First World War, and the intensely nationalist eugenic efforts to populate the continent with desirable Britons, a question still haunted public health policy-makers, plantation owners and investors, government bureaucrats and ministers, academics in geography, microbiology and physiology: 'Is White Australia Possible?'[8]

Much of this chapter focuses on the interwar period, the years in which discourses of degeneracy and nationalism dovetailed as eugenic organisations and ideas, and entered broader English-speaking culture as a eugenic mentalité.[9] My aim is not only to draw, but also to complicate, the relations between race, whiteness and the eugenic mentalité, and to explore their constant connections with public health lines of hygiene. The shifting meanings and categorisations of 'race', the flexibility between concepts of blood, nation, geography, and civilisation have been richly discussed.[10] Yet the connection between changing ideas of race and racial difference on the one hand, and eugenics on the other is not at all straightforward. Not infrequently these different (if related) ideas are analysed as too seamlessly similar. This has been challenged by several Australian scholars, writing from a national context with extremely complicated historical trajectories of race: explaining the connections and the differentiations between eugenics, the 'management' of Indigenous populations, the exclusions of Chinese, Indians, Pacific Islanders and others, and the vigorous project of whiteness in a settler colony demands a careful historical touch. This is a far more complicated cluster of racial and national problems than Dan Stone addresses in his recent argument for the significance of 'race' in British eugenics.[11] For example, Australian historians Stephen Garton and Russell McGregor suggest that eliding eugenics with policies of assimilation or absorption of Indigenous people into the white population is mistaken, historically and analytically.[12] Developing such nuanced analysis here and in the next chapter with respect to sex, I seek to understand the distinctions between the ideas of racial *difference* in operation in the period – in this case white Australians vis-à-vis Chinese people – and the use of 'race' to signify 'nation' or 'British', whereby 'internal enemies' were problematised eugenically.[13] Specifically, I examine through the Australian case the relatedness, but also the distinctions, between the race-discrimination of the immigration acts (and their health rationales) which was their initial driving purpose, and the subsequent eugenic ambitions in screening out healthy from unhealthy whites, which came to be their effect, especially in interwar implementation. The interchangeability (or not) of 'race' and 'nation' has also drawn considerable historical attention in the European context. Race was not a synonym for nation, argues Stone, 'unless one accepts that the word "nation" itself carried implicit racist assumptions'.[14] But this is precisely what 'nation' in Australia did mean, and not just implicitly, but explicitly in terms of law, culture, science and politics. This is why the white settler colony-turned-White Australia is critical for recent scholarship on eugenics, nationalism and race: whiteness *was* the national identity of this particular twentieth century imagined community.

International hygiene

In the last chapter I examined the growing international systems governing the global movement of ships in the nineteenth century: the quarantining measures, the bills of health, the certificates of free movement. Beginning largely as methods for the regulation and inspection of *vessels*, these inter/national systems increasingly came to govern the movement, restriction, identification and inspection of *people*. As I have discussed, they became an integral part of national and international means of identification and surveillance, part of, indeed 'the invention of the passport'.[15] The sanitary conferences of the nineteenth century, attempting to standardise quarantine, grew into national systems of medico-legal border control which shaped populations in terms of race and health. By the interwar period, the need for international communication and agreement on public health issues developed into various international health organisations, for whom quarantine was but one aspect of new attempts at a global administration of life. There were increasingly complicated and criss-crossing communication networks between different configurations of national and colonial powers: the imperial communication on disease and epidemic from British colonies to the Colonial Office, the League of Nations' Health Committee networks, the Pan-American Sanitary Bureau, the Pan-Pacific network involving the Rockefeller Foundation.[16] Like national public health systems that governed contact and movement, in the twentieth-century field of international hygiene, whatever circulated globally was considered dangerous. For example, the problems of opium traffic as well as the 'traffic in women and girls' were as much hygiene problems as anything else. International Hygiene agreements also regulated and discussed the movement of merchant seamen who constantly travelled and connected the maritime globe, spreading venereal and others diseases. 'The Health of Seamen', for example, was one of C.W. Hutt's chapters in his *International Hygiene*, along with 'The Hygiene of Emigration and Immigration', 'Venereal Disease', and 'Land Frontiers and Infectious Disease'.[17] The International Hygiene bodies created new classification systems including the International List of Causes of Death, International Code of Zoological Nomenclature, and a set of internationally agreed 'Signals'.[18] But the central problem of International Hygiene was the global movement of immigration and emigration, its micro- and macro-biological effects, and the various national and international systems for its regulation. This was all the international biopolitics of the twentieth century.

Originally under the rubric of 'race', and increasingly under the broader (but still racialised) rubric of un/fitness, people were newly classified and excluded by law in many national and colonial contexts between the 1880s and the Second World War.[19] If the cholera epidemics drove much of the nineteenth century international quarantine measures, the sudden concern about leprosy and especially its connection with the Chinese diaspora from about the 1880s coincided with the new (but related) immigration and emigration restrictions which appeared classically in the Australian colonies, but also in many other contexts.[20] This 'hygiene of emigration and immigration', as C.W. Hutt put it, was about both public health and changing configurations of race.[21]

There was, then, a sudden rush of race-based immigration restriction in the 1880s. A Canadian Immigration Act of 1885 imposed a restrictive head tax on Chinese people and a more exclusive Act was passed in 1923. In 1881 New Zealand brought in a Chinese Immigration Restriction Act, with a further Act of 1908. Also in 1881 fairly uniform Chinese exclusion acts were in place in New South Wales, Tasmania, Queensland, and South Australia. By 1888 the self-governing colony of Western Australia also had a severely restrictive Act.[22] Likewise in the United States there was statutory exclusion of Chinese people specifically from 1882.[23] Partly because of these new exclusionary acts appearing in the self-governing colonies of white settlement, the Imperial Government in London began to express caution about prohibiting Chinese and especially Japanese people from the space of the British Empire. As the one time governor of Hong Kong John Pope Hennessy put it, 'the common opinion in Parliament is that the governing classes in China as well as the people of China would be offended if we prohibited Chinese immigration into a British colony'.[24] British concerns were about the security of commercial relations with China, and the viability of what were by then intricate networks of especially Chinese contract labour throughout the Empire.[25] Possibly more importantly, the Imperial British government was concerned with the security of military and diplomatic relations with Japan. And finally, there was the perennial issue about the legitimacy of limiting the movement *within* the Empire of 'coloured' subjects *of* the Empire – in particular Indians.[26]

As we shall see, Imperial anxieties about racial exclusion strongly shaped Australian legislation, but the general discussion about race, exclusion and the concept of racial equality went to another level altogether after the First World War. It was Australia's exclusionary law, the Immigration Restriction Act of 1901, which was essentially the test case at the Peace Conference. The Japanese delegation strongly opposed the principle and the practice of race-based exclusionary law, seeking the inclusion of the Racial Equality Clause into the covenant of the League of Nations.[27] Under major debate was the whole question of whether a sovereign nation should have the right to exclude certain people on the basis of race, or to put it positively, to determine its future racial development by enacting such immigration laws as it saw fit, and as a solely domestic concern.[28] The 'international' decision to permit sovereign determination of racial exclusion was a difficult one, and not just because the Japanese view was so strongly represented. The concern was also the longstanding one especially for the British Imperial government, that exclusionary laws opposed principles of liberty of movement. As J.W. Gregory, Professor of Geology at the University of Glasgow put it in 1928, 'the right of any person to move from one country to another has been widely regarded as indispensable to personal liberty'.[29]

A trend is evident internationally, from the explicit nomination of nationality or race in the exclusionary acts of the 1880s (in which health, hygiene or a threat of contagion is implicit), towards other criteria of 'undesirability', including health and hygiene criteria, in which *race* is implicit. In other words, explicit racial categorisation of the late nineteenth century was often declined by later legislators, and exchanged for implicitly racialised categories of infectiousness or unfitness, or for technical means of exclusion without specifying race. For example, from 1910 Canadian immigration agents were able to exclude people determined to be 'unsuitable for the climate' – that is, Afro-Caribbeans largely from the United States or the West Indies. They might also be excluded as 'undesirable owing to their peculiar customs or modes of life, or because of their probable inability to become readily assimilated, or to assume the duties and responsibilities of Canadian citizenship'.[30] In South Africa the Immigration Restriction Act of 1913 also drew on unsuited 'standards or habits of life' as the flexible device of exclusion.[31] Australian legislators could hardly use the 'climate' option, given the problematisation of the tropical climate for the 'white man' – as we shall see, a considerable difficulty in itself. Rather, they used the idea of a dictation test in any European language, in which an immigration (or as it turns out, quarantine) officer would specifically choose a language *not* that of the prospective immigrant.[32] This 'dictation' or 'education test' had already been used in prior immigration restriction acts in the colonies of Western Australia, Tasmania and New South Wales as well as in Natal.[33] The New Zealand immigration restriction act also employed the 'dictation test' mechanism, but interestingly did so with the use of a negative race identification, the specification of any person *not* British-white as requiring the test: 'Any person other than of British or Irish birth and parentage who, when asked so to do by an officer appointed under this Act by the Governor, fails to himself write out and sign, in any European language, an application' was deemed prohibited.[34] A broad move away from racial specificity is evident in nearly all of these examples. As explicitly race-based exclusionary laws became more difficult to sustain – for whatever reason – discriminatory practices were constituted in coded ways (assimilability, climate and constitution, habits of life) as well as technically evasive ways (the need for a continuous voyage, the impossible dictation test).

By the interwar period, international migration was comprehended – and regulated – as a biological issue. Under eugenics and early genetics, the social body of early to mid twentieth century states was imagined much more literally (that is to say biologically) than had been the case in the nineteenth century. Indeed the connections between migration, population, eugenics and genetics were to continue through much of the twentieth century.[35] The regulation of immigration, like public health, was a biopolitical measure, a measure in which both the quantity and nature of the population was managed in a calculated way. But it was done so, as both national and international biopolitics. Because immigration shaped populations, sometimes in very marked ways, and created through reproduction the possibilities for better or worse, more or less fit populations in the future, policies about the movement of people within the Empire and Commonwealth, as elsewhere, came to be considered within a eugenic logic. Eugenic sensibilities as well as more formalised eugenic programmes contributed to this biological constitution of migration in the interwar period, re-imagined in terms of the loss or gain of good or bad 'stock' into differently imagined and bounded populations. As W.E. Agar, Professor of Zoology at the University of Melbourne put it in 1928, 'The future population of Australia will be derived from two sources – from the descendents of those already here, and from immigration'.[36] And in Britain, 'alien immigration' was similarly considered undesirable by the Eugenics Society for two reasons: '1) as altering the racial type by the influx of foreigners who in other respects are not undesirable persons, and 2) as introducing persons, who, quite apart from their race are undesirable'.[37]

Here the two faces of 'race' are evident: the concern about colour and racial difference, and the subsequent concern about 'internal enemies' within the nation or within the 'white race'. Internal enemies were those who carried bodily conditions which could spread – communicable diseases like tuberculosis, syphilis and gonorrhoea. But also in this period, there was a biologising of previously moral states and tendencies – criminality, homosexuality, imbecility. That is, these conditions could 'infect' the next generation, a point I extend in the next chapter. Moreover, these years saw increasingly detailed distinctions between categories and criteria of mental health and hygiene. In Australia, Canada, the United States as well in Britain itself under its Aliens Order of 1920, mental health and hygiene criteria were ever more finely written into immigration law and regulation.[38]

The eugenic discussion of migration imagined different kinds of bounded and segregated populations. For British organisations, the bounded population at issue was sometimes Britain itself, as the fit moved to other parts of the Empire (especially Australia and Canada), or as the 'unfit' were repatriated home from the Dominions. At other times the bounded population at issue was trans-national – the white race within the Empire itself. Eugenicists and public health authorities sometimes sought to implement policies which safeguarded the quality of whites within the heterogenous racial population which was the twentieth century British Empire. The 'problem' of coloured British subjects' right of movement which had so troubled the Canadian, Australian, New Zealand, and South African governments in the late nineteenth and early twentieth centuries as they implemented various exclusion acts, was now refigured squarely as a eugenic issue. The London-based Eugenic Education Society, established in 1907,[39] objected to the right of any British subject to move freely within the Empire, especially into Britain: although 'British' subjects, they were not necessarily of the same race. The Society was concerned that 'only nationality' and not 'race' was recorded on official documents. Thus, they argued, people became officially 'British persons who are racially foreigners'.[40] The distribution of the white population within the Empire was one issue which the British Eugenics Education Society took up directly with emigration authorities in Britain. It both sought information from, and wanted to advise Dominion governments on, three main issues: inter-marriage between races; the effects of climate on fertility; and 'the home origin, and subsequent distribution, of emigrants to the Overseas Dominions'. As it was put in one letter to the 'Colonial Premiers' in 'all the great self-governing countries which in confederation constitute our Empire' these three major issues affected 'the quality of the population of the Empire as a whole'.[41]

The movement of populations over borders was a biological issue which made it both a public health matter in this period, and a eugenic matter. The problem of 'population' in Australia and in imperial Britain was not only about the regulation of reproduction but also the biopolitical management of migration. This was problematised and implemented within imagined geographies which were sometimes national, sometimes imperial, and sometimes the racial interior frontier of Empire which bounded white from non-white. Moreover, as international biopolitics, this hygiene did not function only at the level of national and international relations, but increasingly governed individuals: their movement, their sense of national citizenship and, through growing bureaucratic systems of documentation, their official 'identity' as British Subject, Foreign National or Coloured Alien, for example. But national or raced identity was heavily circumscribed by bodily/health classification and documentation. Thus, as I have shown in previous chapters, people were also vaccinated or unvaccinated, carrying tuberculosis or never exposed, from a clean or a diseased region, mentally sound or feeble-minded. Classification (and subjectification) then, were not only presented in terms of nationality and race but in terms of cleanliness, fitness and health.

Racial imaginings and white Australia

Australia's Immigration Restriction Act (from 1912, the Immigration Act) was exemplary in respect of this international trend from explicit to implicit racial reference in immigration law and policy, from a discourse of racial difference to eugenic discourse on the (moral, physical and mental) hygiene of whites. Indeed it might be seen as the hinging Act of the period. Passed in 1901, it inherited the spirit and the intention of the race-based exclusions of the 1880s and 90s, but it was also a foundational Act for the new 'eugenic century' in its subsequent use as a way of discriminating between whites.

The idea of white Australia was developed and entrenched around the problematisation of Chinese men in the colonies. Many Chinese people had arrived in the Australasian colonies with the gold rushes in Victoria and New South Wales from the 1850s, and in Queensland from the 1870s. Entrants were almost exclusively single Chinese men: for example in New South Wales in 1878 there were 9,616 Chinese (181 of whom, incidentally were married to white women) in a much larger populations of whites. In 1888 in the Northern Territory (at that point part of the colony of South Australia) however, there were about 7,000 Chinese and fewer than 1,000 whites.[42] By the 1880s, particularly in Queensland, virulent anti-Chinese agitation centred on questions of labour and unwanted competition, and the trade unions and the new Labor Party itself validated the anti-Chinese stance. 'Australia for the White Man' was a maxim which expressed masculinist working-class identity, national identity and racial identity simultaneously.[43] Sir Henry Parkes, Premier of New South Wales, offered many statements on Chinese immigration, one speech summarising widespread ideas – his speech on the Influx of Chinese Restriction Bill in 1888:

> I contend that if this young nation is to maintain the fabric of its liberties unassailed and unimpaired, it cannot admit into its population any element that of necessity must be of an inferior nature and character ... we should not encourage or admit amongst us any class of person whatever whom we are not prepared to advance to all our franchises, to all our privileges as citizens, and all our social rights, including the right of marriage.[44]

These comments illustrate the early nationalism which anti-Chinese sentiment encouraged, the way in which early expressions of 'white Australia' were about whiteness and Australian nationalism, as well as a thoroughly British cultural, social and racial identity determined through freedoms and liberties, through civic rights and duties.

A significant articulation of this racial politics was bodily and biomedical; a language of contamination which conflated the moral and physical. In 1901, for example, the Labor Party leader J.C. Watson argued that 'the objection I have to the mixing of the coloured people with the white people of Australia ... lies in the main in the spoilibility and probability of racial contamination'.[45] Isaac Isaacs, later Governor General and High Court Justice, spoke of 'the contamination and the degrading influence of inferior races'.[46] There were two dominant images of Chinese men which became widespread from the 1880s: first, as sexual and moral threat to white women, thus a threat to national integrity; and second, as purveyors of disease. These were to some extent contested images.[47] But over and over again in dominant representations of the future of the colonies and the new nation, fear of unrestricted immigration and even of invasion was enunciated as fear of disease, of smallpox and especially, as we have seen, of leprosy.

Many historians have detailed the history of anti-Chinese activity, law and policy, the difficulties of indentured Pacific labourers, and the relation between the white Australia of immigration restriction and policies aiming to manage Indigenous populations.[48] My concern is specifically the exploration of the ways in which this dovetailed with questions of health and infectious disease control, in particular quarantine and international hygiene, and what this meant for the imagining and technical implementation of Australian 'whiteness'. But two aspects of racial management in Australia are important to note first. While there were other policies which managed Indigenous people through health measures some of which I have examined, the 'white Australia policy' was considered at the time a separate policy dealing specifically with immigration and naturalisation. As it was put in the *The Empire Review* in 1909: 'it was not intended, of course, to operate against the native aborigines, but these are rapidly and naturally dying out'.[49] This dominant idea of the decline of Aboriginal populations was problematic in all kinds of ways,[50] but it is important to recognise that many white individuals and groups, as well as experts and government distinguished quite clearly between the idea of 'white Australia' and the 'Aboriginal problem'.[51] Second, exclusion on the basis of race was never total: indeed, like the quarantine regulations, immigration regulation was 'leaky' in practice, sometimes intentionally sometimes unintentionally. Numerous Chinese people stayed in the country after the various restriction acts, others came and went, 'slipping through the system'.[52] And many Pacific Islander communities both chose to and eventually managed to stay in northern New South Wales and Queensland in the early twentieth century.[53] Officially, certificates of exemption from the dictation test could be and were granted, for example to Chinese merchants and traders or Indian students or Japanese pearlers. There never was, then, a 'white Australia'. Nonetheless, it was almost universally believed by Europeans in Australia that the new nation *should* be 'white', and must not be 'Chinese'.[54]

Like the desire for uniformity of quarantine measures across the colonies, an 1888 Intercolonial Conference successfully exerted pressure on colonial governments to create uniform immigration restriction law and policies.[55] The conference was led by Sir Henry Parkes who in the 1881 smallpox epidemic in Sydney had tried to exclude Chinese *en masse* through quarantine measures.[56] At an 1896 Intercolonial Conference, it was agreed that the colonies should extend their anti-Chinese legislative measures to incorporate all so-called coloured races, and that the exemptions for 'coloured' British subjects should be withdrawn.[57] Initial drafts of a federal Australian immigration restriction act specified race and colour. But a series of Imperial secretaries and ministers opposed this. As contemporary commentator and geographer Persia Crawford Campbell put it, 'the only check on their [the Australian government's] determined action being the interference of the Imperial Government'.[58] The Colonial Secretary Joseph Chamberlain indeed interfered strongly and successfully in 1897, indicating that immigration restriction should be legalised but only 'without placing a stigma upon any of Her Majesty's subjects on the sole ground of race or colour'.[59] All this meant that the Australian act never mentioned race.[60] But it did not need to: by 1901 the rhetoric and politics of 'White Australia', one of the world's most resilient national slogans, was widely recognised, nationally and internationally.

On a surface reading of the Immigration Restriction Act, discrimination on some sort of biological or bodily definition of race was rejected. But of course at the level of implementation, the assessment of race by government officers was precisely a bodily/visual mode of discrimination between potential immigrants: it was on those unstated grounds that the dictation test was ordered. But there was another mode by which the biological and the bodily were foregrounded in the working of Act, and one which has been obscured by scholarly interest in the device of the dictation test. Written into it were public health rationales for the prohibition of entry of any individual. In addition to the exclusion of 'any idiot or insane person', criminals and prostitutes, Section 3d of the Act named as a prohibited immigrant 'any person suffering from an infectious or contagious disease of a loathsome or dangerous character'.[61] Immigration regulations then, were quite literally health regulations.

The reverse was also the case in this period. Cumpston, as we have seen, nominated quarantine as the premier strategy in government pursuit of 'national cleanliness' and he did this specifically in racial terms. The whole object of quarantine, he wrote, 'is the keeping of our continent free from certain deadly diseases at present unknown amongst us. And secondly, the strict prohibition against the entrance into our country of certain races of aliens whose uncleanly customs and absolute lack of sanitary conscience form a standing menace to the health of any community'.[62] One of the designated functions of the Division of Marine Hygiene of the Department of Health discussed in the last chapter, was the medical inspection of passengers and crew under the Immigration Act. Conversely, it was understood that the public health grounds of the Immigration Act would be used to remove from Australia people with longer term infectious diseases, like venereal diseases or tuberculosis, which were not, or could not, be ascertained at quarantine inspection. It was stated that 'The responsibility of the Commonwealth for the exclusion of communicable diseases from Australia or from any clean State may ... reasonably be recognised under the Quarantine and Immigration Restriction powers'.[63] In many ways these were approached as twin statutes. The statutes which delimited and defined the practices of quarantine and the practices of immigration restriction explicitly referred to, and required, one another.[64]

One can see this at work in those few sites on the edges of the nation where 'coloured' labour *was* permitted, where exemptions to the Immigration Act were made. At Thursday Island in the north and at Broome in the west, such exemptions were common in the interests of the pearling industry which relied on indentured labour. Since the immigration regulations were loosened, the health powers were tightened. As Cumpston put it in a memo to the Department of Home and Territories, 'amongst the conditions under which indentured coloured labor is permitted is one to the effect that medical certificates, showing that the laborer is free from disease, are to be produced'.[65] Here, the Empire was bureaucratically at work. Those labourers arriving from colonies or territories of the British Empire – Singapore or Hong Kong for example – were supplied with certificates of freedom from disease and fitness for work from medical officers in those ports. These certificates accompanied their contract for labour. For those arriving from Koepang, Macassar, or Japan, or places where there was no medical officer, the quarantine officer would subject the indent to 'a very critical examination before being allowed [to] land'.[66] Evident here is the relation between immigration regulation and health regulation, the clear sense in which both statutorily and in practice, these were twin and complementary regulatory mechanisms.

One journalist wrote of Cumpston's powers thus: 'He controls the quarantine service and conducts a ceaseless war against the foreign germ declared by his department to be a prohibited immigrant'.[67] Here the germ and the immigrant were conflated. Quarantine dealt with both, and immigration restriction dealt with both; there was a range of ways in which they were conceptualised as the same thing. But precisely because the Immigration Act was implemented to exclude the 'coloured alien' from the territory and the body politic of Australia, the real work of this medico-legal border control was undertaken on those who *were* seeking entry, and indeed who were sought by Australian governments – whites.

Imperial migration and racial hygiene

The racialised aspiration of white Australia was only partly about processes of exclusion. Although the conflation of disease especially with Chineseness at the turn of the century clearly came into play in both quarantine and immigration restriction, the restrictive Acts themselves meant that thereafter the implementation of these policies was about monitoring and policing the health of Britons seeking entry: it was primarily a eugenic ambition of public health. And one of the rationales for the exclusion of would-be immigrants was the exclusion of the 'undesirable' on medical grounds: the infected, the mentally unsound, those with contagious 'tendencies'. While the original rationale of the Australian Immigration Restriction Act was to exclude coloured aliens, it was also, and in an increasingly refined way, a mechanism for screening white entrants through what might be thought of as the mental hygiene and the physical hygiene sections of the Act. As they were implemented over the first half of the twentieth century, then, the medico-legal border systems of both quarantine and immigration in the new Australia were largely implemented with respect to the moral and physical quality of aspiring British (and other white) entrants.[68] In practice, it was not 'coloured aliens' who were subject to medico-legal border control, because they were refused entry under other sections, but rather white-British people who were subject to the increasingly detailed procedures: medical examinations, mental health tests, compulsory vaccination, genito-urinary examination for venereal diseases and later chest x-rays for tuberculosis.

In the nineteenth century colonial period, systems of inspection were constantly changing and were haphazard. Non-assisted immigrants seldom required medical examination at point of departure in Britain, but assisted immigrants did. With varying levels of compulsion then, a certificate of health from any doctor needed to be presented to the Agent-General of the Colonies in London, in order to secure passage. This needed to be

> signed by the Medical Officer of Health for the district from which they come declaring the state of the Public Health in that district as to infectious disease ... if small-pox is epidemic in the district from which any emigrant comes every such emigrant shall be vaccinated or re-vaccinated as the case may be before he shall be allowed to embark.[69]

Significantly, these 'inspections' were less concerned with the actual state of health of the person, than a statement about the presence or absence of infectious disease in the area from which they originated. This was to change to a more individualised surveillance, where health identity and personal identity merged. By 1913, for example, not only medical officers at point of departure, but ships' masters and medical officers were asked to certify the health or ill-health of each individual passenger, indicating 'whether he or she is insane or medically defective ... suffering from epilepsy, pulmonary tuberculosis, trachoma, or any loathsome or dangerous communicable disease'.[70]

There was a series of amendments to the Australian Immigration Act in 1912 which reflected and enforced a growing bureaucratic structure surrounding movement between nations. The initial general prohibition against 'idiots and insane people' became 'any idiot, imbecile, feeble-minded person or epileptic'. A general prohibition against those with loathsome or contagious diseases became 'any person suffering from a serious transmissible disease or defect; any person suffering from pulmonary tuberculosis, trachoma, or with any loathsome or dangerous communicable disease, either general or local'. Moreover, entry thereafter positively required a prescribed and standardised certificate.[71] Also in 1912, a Commonwealth Medical Bureau was provided for to streamline the haphazard procedures of the old colonies (now the States) and to add a Commonwealth of Australia governmental presence, not just a State governmental presence in London.[72] Notwithstanding the status of a person's health certificate, under Quarantine powers any immigrant, crew member or visitor could be medically examined, detained and/or vaccinated on arrival at an Australian port for any disease or condition nominated under the Quarantine Act or the Immigration Act.[73] These did change periodically, but in 1915, for example, the infectious diseases specifically nominated were smallpox, plague, cholera, yellow fever, typhus fever, leprosy, anterior poliomyelitis, cerebro-spinal meningitis, malta fever, scarlet fever, chickenpox, measles, whooping cough, gastroenteritis, typhoid fever, diphtheria, malarial fever, gonorrhoea and syphilis.[74] In 1917 tuberculosis was added, in line with the Immigration Act, as well as soft chancre and venereal bubo because of the increased awareness of venereal diseases in the War.[75] And in 1919, separate regulations altogether were enacted in relation to the influenza pandemic then sweeping the world.[76]

The logic of restricting entry was in the first instance a public health quarantining one: a means to stop the entry or spread of communicable diseases. But, especially for the chronic diseases of tuberculosis, syphilis or gonorrhoea, as well as the moral and mental criteria, the logic was eugenic. The imperative of white Australia was effected as much by a eugenic screening out of 'undesirable' Britons, as the more comprehensive screening out of various Asian individuals and populations on the basis of race. In this sense, the 'whiteness' of white Australia was not only a whiteness understood through racial difference, but also a whiteness understood and sought, eugenically. The ideas and activities of the Racial Hygiene Association of New South Wales illustrate this well. They show the almost complete identification of race and nation in Australia in this period, and the ways in which the ambition of racial hygiene drew immigration into direct consideration as a site of public health governance.

Originating as a feminist organisation (and becoming the Family Planning Association),[77] this group was the equivalent of the British Social Hygiene Council: their driving problem was the social and educative prevention of venereal diseases. For such associations and for the related industry in birth control, the interchangeable use of 'race' and 'nation' was common. So when Marie Stopes dedicated her *Wise Parenthood* 'to all who wish to see our race grow in strength and beauty', she meant British society or the nation.[78] Yet in the Australian context and for the Racial Hygiene Association, race and nation were not just equivalent, they defined each other as political projects: they had a slogan – white Australia.[79] For the Racial Hygiene Association, 'national hygiene' always presumed racial specificity because of the exclusionary aspect of the Immigration Restriction Act: 'the national health of *whites*' was the presumed understanding. But to read 'racial health' as securing white identity *only vis-à-vis* an Asian other which threatens its purity, is to miss the point. This may have been the case in the 1880s and '90s when white Australia was first being promoted. But by the interwar period, the (imagined) racial homogeneity of Australia was a cultural and political given. In its own terms, racial hygiene was about whitening and purifying Australia from the contaminations of other (but tarnished) whites. This was the case not only with respect to those 'native' to, or naturalised within the territory and social body of Australia,[80] but also, even especially, those Britons seeking entry whose health threatened to diminish rather than enhance the quality of whiteness of the nation. 'Race and health' (or 'racial hygiene' for this Association) meant intervening in the quality of national whiteness.

I take up racial hygiene in terms of venereal disease and reproduction in the final chapter, but here I am interested in this group's engagement with immigration. Importantly for my purposes, the Racial Hygiene Association saw immigration as a health issue because it was a 'race' issue. This conflation is apparent in a report the organisation submitted to the Prime Minister in 1928, 'A Report on

Immigration (as affecting Racial Values and Public Health in NSW)'. 'Immigration is primarily a racial matter,' they argued. They did not mean for example that the Chinese might bring leprosy or smallpox into the white race/nation of Australia, as the arguments had run in the 1880s and '90s. Rather, they meant squarely that more stringent health screening procedures needed to be instituted, to deter and turn away mentally, physically and therefore morally unfit Britons.[81] The report delineated in a telling way between the category of 'Australian' and 'non-Australian', offering statistics on the comparative levels of healthiness (variously venereal diseases, tuberculosis, insanity, criminal tendencies) between the two groups within Australian institutions (prisons, asylums, infectious disease hospitals). Significantly, the categorical distinction at issue was between those born in Australia on the one hand, and those born in 'England, Scotland, Ireland, Wales, or Foreign Countries' on the other.[82] Here, 'Australian' meant whites in Australia and 'non-Australian' meant British or foreign whites, not coloured aliens. The problem, according to the authors was that the health of the national (white) population was diminishing due to the unwanted migration of the 'residue' of the British population (the fit being attracted to North America) and due to inadequate health screening and standard of examination of Britons at points of departure as well as entry points – the quarantine and immigration procedures on reaching Australia. In this instance, then, regulating inclusion (of healthy and fit Britons) not exclusion (of coloured aliens) was the concern of the Association: checking in London and in British ports for tuberculosis, venereal diseases, and mental health. Eugenically unfit whites were the internal enemies of the biopolitical state in a way which was premised on the racial exclusivity of 'white Australia', but was not identical to it. The stakes were high, according the self-important Racial Hygiene Association of New South Wales: 'Australia is the only continent practically free from colour problems. We hold in our hands the opportunity to become the headquarters of the White Race, and the centre of civilization'.[83]

Needless to say, this was a deluded ambition from the perspective of many London commentators, not to mention indigenous activists. There was a long nineteenth century British tradition of imagining the colonies of 'settlement' as release-places, empty places where the pressures of industrialisation – the 'sanitary' problems of space, population and resources, producing poverty, sickness and immorality – could be relieved. In the international hygiene literature of the interwar period, the Australian continent was still empty.[84] This was an extremely powerful idea, premised on and sustaining a cultural and legal obliteration of Indigenous people's presence and claim to land, as well as shaping tropical medicine debates about who should fill this emptiness, and how. Population literature and medical geography often worked through hydraulic metaphors in which fluid populations moved through unstoppable natural laws. For example, Woodruff's *Expansion of Races* (1909) was conceptually structured through this metaphor. 'Population as a Fluid' was explained in his first chapter. Subsequent chapters dealt with 'Movements in Confined Fluids and Populations', 'Wave Motions' and 'Saturation Point of Populations'.[85] Population and hygiene literature was extensively interested in the restrictive migration regulations that had emerged from the late nineteenth century, and the sense in which these had built up international tensions: segregative barriers were 'dams' which may not be able to, and possibly should not seek to, artificially contain and hold back 'natural' global population movement. International commentators not infrequently thought the empty Australian continent was unsustainable in this respect. It was even argued that the northern portions should be 'given to' the Japanese to relieve international population pressures, and therefore political pressures, as the Dean of Canterbury controversially suggested in 1933.[86]

'Emptiness' also governed Australian official imperatives to populate. Reproduction of the white race already within the continent was not going to solve this problem of emptiness quickly enough: migration was the solution. But governments and groups like the Racial Hygiene Association were continually caught between wanting to facilitate movement of white-Britons and various kinds of Europeans into the continent, but at the same time avoid thereby introducing into the body politic the 'internal enemies' of the eugenically unfit. For example, the Racial Hygiene Association was deeply troubled by plans made jointly between the Australian and British governments by which people in various pauper institutions in England would have their passages paid. This might be desirable for British authorities and help fix the Australia population problem numerically, but it would raise far greater problems in terms of the health of the future Australian population.[87] This is where the dual governmental barriers of immigration and quarantine needed to be strengthened, it was argued: one line of defence at the medical examinations in London, the second line of defence at the national quarantine border, now a line of eugenic screening, letting some in, keeping others out. Between Britain and the Dominions there was an ongoing interwar discussion about which nations were gaining and losing the valuable fitter types and the undesirable unfit.[88] If the Racial Hygiene Association and the Australian governments were concerned about receiving 'bad stock' from Britain, British commentators were more concerned with the reverse problem: their fittest being drawn to the Dominions, including Australia. Fleetwood Chidell's book, for example, *Australia – white or yellow?* (1926) was in large part taken up with the biological ramifications of emigration on Britain. 'Expatriation of the best' he called it.[89] And J.W. Gregory wrote that 'Some British writers complain that Australia is too fastidious and rejects far too many of those who are anxious to settle in Australia. The complaint is that Australia will only receive the very pick of the British workers and is unwilling to receive a fair share of the unemployed who are now crowded into the British towns'.[90]

No matter how insistently Australian governments sought fit British whites, however, there remained the co-incident and deeply related problem of 'the white man in the tropics'. If the various exclusion acts had marked Chinese, Japanese or Islanders as foreign, the problem of the tropics marked the white body as a foreign body just as forcefully. In Australia, the problem of 'the white man in the tropics' was taken up directly as a problem of public health, government and medicine, and was an immediate effect of the exclusionary immigration acts.

Tropical medicine and foreign white bodies: 'Is White Australia Possible?'

Tropical medicine was a discourse which arose out of longstanding ideas about the foreignness of white bodies in the tropics: it was thus part of the imperial project.[91] The idea that there was a limit on the possibility of permanent and healthy European settlement in the tropics became deeply entrenched. To be sure, this was not unfounded as European military as well as commercial experiences of the tropics from the early modern period were often disastrous in terms of morbidity and mortality.[92] The interest of European medicine in diseases of warm climates and in tropical medicine closely followed the various imperial and economic interests around the globe: British in the West Indies and the Straits; Dutch in the East Indies; later the French in North Africa; Americans in the Philippines; Australians in Queensland and New Guinea. All of these colonial engagements were shaped and accompanied by a network of medical research and administrative effort, institutionally and intellectually defined through the fields of 'diseases of warm climates' and its successor, tropical medicine and hygiene. While new microbial studies to some extent differentiated the latter from the former, sustained through both was a dominant aetiology of health and ill-health that geography, temperature and climate were causative of disease in European as well as non-European constitutions. This was why place was so significant, why medicine and health were so connected to the imperial project and why geography, migration and hygiene were coinciding intellectual and practical specialties.[93]

The discipline of tropical medicine in Australia in the early to mid twentieth century was shaped and driven officially by the nationalist question: 'Is White Australia Possible?' In this period of race-based nation formation and the implementation of the various exclusionary immigration acts, the biological nature, capacities, and possibilities of whiteness came to be defined and investigated. The nature and qualities of whiteness itself were always and firmly in biomedical site, literally under the microscope. Indeed this was a site which strangely reversed the more usual gendered and raced dynamics of modern western medico-scientific research. Rather than studying the black body it studied the white body, and, as I have argued elsewhere, rather than pathologising women, it pathologised men.[94]

The Queensland sugar industry had relied heavily on the labour of indentured Pacific Islanders. Thus the turn-of-the-century outlawing of 'coloured labour' was a twofold problem. Not only did the exclusion and deportation Acts essentially remove a labour force (not totally, but certainly extensively) but this removal also left the residual problem, the question whether 'white man' could even *live* sustainably in the tropics, let alone undertake the heavy physical labour of the plantations.[95] As Australian tropical medicine experts like Raphael Cilento argued, there was a pressing economic imperative that white labour fill, and be seen to be *able* to fill, this industrial gap. There was also, as we have seen, a strong nationalist imperative to populate the northern edges of the new nation, to be seen to utilise and cultivate it, that is, to own it. In their own terms, and in concert with the idea of tropical medicine, the securing and segregating of 'white Australia' through the immigration and deportation acts posed as many problems as it solved.

The link between the imperative of white Australia and tropical medicine was explicit in the bureaucratic development of health, and in early twentieth century scientific and medical texts.[96] As Cumpston put it: 'One of the most cherished of Australian ideals is the preservation of this continent for occupation by those racial stocks which have developed it to its present condition. In view of the large proportion of its total area which lies within the tropics, important considerations of tropical hygiene and of physiological adaptation to tropical temperatures and humidity have had to be considered'.[97] And in one of many direct links drawn between tropical medicine and the Immigration Restriction Act, he explained:

> The very extensive discussion inevitably associated with the adoption, in 1901, of the 'white Australia' policy brought sharply under review the conditions of life in the tropical regions of the Commonwealth. The decision to cease the importation of Polynesians to work in the sugar plantations, and to repatriate those already in Australia, raised the issue whether white Labour could survive hard work under tropical conditions.[98]

The official brief of the government-funded Australian Institute for Tropical Medicine in Townsville was the investigation of the possibility of permanent white settlement in the North. Anton Breinl, from 1910 the first director of the Institute defined Australian tropical medicine thus: 'The object of tropical medicine involves much more than the study of parasites and diseases occurring in the tropics; it comprises in its working sphere the welfare and life of the white man under new, and, to him, artificial conditions'.[99]

'Is White Australia Possible' was the structuring research question of the local discipline of tropical medicine. The publications emerging out of this research industry were prolific: J.S.C. Elkington's *Tropical Australia: Is it Suitable for a Working White Race?* (1905); W.J. Young's 'The Metabolism of White Race Living in the Tropics' (1915); Raphael Cilento's *The White Man in the Tropics* (1925), and 'Observations on the White Working Population of Tropical Queensland' (1926), to name a few. Australian medical professional bodies approached the question very seriously indeed. In 1920 the Australasian Medical Congress met in Brisbane and took the medico-political issue of the white race and the tropics as its theme. The proceedings were published by Government order, compiled by Cumpston and titled 'Tropical Australia'.[100] Several esteemed international men of science became deeply engaged in the problem of the Australian tropical environment, its peopling and its 'uses', for example the Yale geographer Ellsworth Huntington and the Sydney-trained international geographer, T. Griffith Taylor.[101]

Australian tropical medicine experts interrogated the white body biologically to its most minute levels. The physiological, biochemical and microbiological studies were done primarily on white men, sometimes on white children, but less commonly on white women. Much of the early work at the Institute in Townsville, for example, was a series of experiments on aspects of white skin. A number of studies problematised white skin in terms of an idea of white 'complexion'. This stemmed from a notion of 'tropical anaemia' in which the same person who was 'rosy' in cold climates had a certain pallor in the tropics. Breinl and Young explained in 1919: 'The skin of the healthy European inhabiting a tropical climate appears to the newcomer pale and sallow, the degree of the sallowness depending on many factors, especially complexion and skin texture ... The paleness of the skin has naturally given rise to the conception that there exists a tropical anaemia and that "thinness and poorness" of blood is natural sequence of prolonged residence in the tropics'. This, they concluded is 'from the scientific point of view only a myth'.[102] Of concern also was the effect of the sun in turning white skin darker. For example, the Melbourne doctor J.W. Barrett asked in 1925: 'Can heat, and heat alone cause physical demarcation?' That is, will whites living in the tropics turn brown? 'We know' he answered, 'that however sunburnt the white races who live in hot climates may become, their children so far, show no signs of inheriting this pigmentation'.[103] Related to studies of white skin and complexion, were studies of blood pressure and blood counts of white men.[104] The question of metabolism was also a central problematic, entailing biochemical study of temperatures and sweat.[105] Men were placed in sweat-boxes, temperatures raised and the quantity and biochemical nature of sweat ascertained.[106] Other studies examined the amount of nitrogen and sulphur excreted in urine, and the contents of white men's faeces.[107] Sundstroem's monumental study *Contributions to Tropical Physiology: with special reference to the adaptation of the white man to the climate of North Queensland*, details the exhaustive range of biochemical analyses of white people.[108]

The discipline of tropical medicine was implicated closely with the federation of the colonies. As we have seen, the north of Australia, and in particular the edges of the north – the boundaries of the nation – were newly significant for the aspiring nationalist culture of racial homogeneity. One of many tracts entitled 'The White Man in the Tropics' explained how tropical medicine, Australian race-nationalism and the international 'hygiene of emigration and immigration' were linked:

> I have been asked to speak on the very important and unsolved problem of whether the white man, and particularly the Nordic

white man, can settle permanently in the tropics, and I need hardly emphasize how vital is this question to the Commonwealth. Again and again other nations have called us 'dogs in the manger' because of our 'White Australia Policy' as regards these empty spaces, and only recently the Dean of Canterbury voiced a very unusual opinion that we should give North Australia to the Japanese.[109]

Such statements worked through, and relied on, the new geopolitical significance of Australia's edges and the interest in distinguishing the whole island from Asian 'others' over the border, persistently constructed as contaminating. But this political and nationalist driving imperative was not always easily accommodated by academic assessment. Not only was there medical concern that whites might not sustainably populate and settle these places, but as both Walker and Anderson have shown, some geographers questioned whether the land and environment itself could sustain large white (or any other) settlement and exploitation, irrespective of the political and race need to secure the north.[110]

As we have seen, some commentators on international hygiene suggested controversially that coloured labour *should* be permitted in Australia, in part as a way of ameliorating international tension over Australian exclusionary laws, and in part as a commercial solution to the problem of plantation labour. In *Australia – White or Yellow*, Fleetwood Chidell advocated this solution, but strictly north of a certain line:

> the international tension would at once be lessened and a comparative friendly feeling would replace the intense dissatisfaction which now only awaits an opportunity to become actively hostile ... Definite frontiers would be drawn which would divide the regions of coloured development from those which were being worked by white men ... Along the border-line the colour bar would be strictly maintained.[111]

This plan never eventuated, but prefigured the Western Australian 'Leper Line' which was in fact a 'colour bar' justified on grounds of public health. Those, like Chidell who argued against the internationally controversial white Australia policy on economic grounds (its damage to Queensland industry, the lack of cultivation of the Northern Territory because of the absence of labour) reversed the tropical problem and relied on the idea that coloured men were naturally fitted to the tropics, and white men were not. This made 'white Australia' not only deeply unnatural but contrary to British principles of fairness. Thus, to claim, but not cultivate or utilise the tropics because of this lack of fit between climate and the white man cut across 'our principles of Imperial rule' as one commentator put it in the *Empire Review* in 1909. These principles, he wrote, 'do not admit that our flag shall exclude entirely the coloured man from regions where Nature intended that he should live and flourish'.[112]

Conclusion

Multiple boundaries of rule were at work in the implementation and imagining of white Australia. The cultural and legal significance of quarantine was useful for a particular racial politics defining and structuring this new Australia. Part of the reason for the centrality of quarantine to public health that I explored in the last chapter was its place in the production of a national 'healthiness' figured racially. 'Purity' was always part of a biomedical/bodily lexicon *and* a racial lexicon, and thus the micro-practices of quarantine doubled as micro-practices of racist immigration restriction. Part of the emerging complex of international hygiene, quarantine and immigration restriction were biopolitical technologies functioning through border control, technologies by which the Australian population was shaped. This shaping took place literally, with the restriction of entry of certain people on grounds of race, and on public health grounds. It also took place as an imagining of the Australian national body as pure but requiring protection, as white, but precariously so.

White Australia had many problems, which the connections with health and hygiene aspirations and practices illustrate. The idea of white Australia was based on racial exclusions legitimated through always more-than-metaphorical concerns about hygiene and contamination, the rendering of coloured aliens as foreign and dangerous in every respect. White Australia was also effected through the complementary project of including particular whites on public health/eugenic criteria. But alongside these aspects of white Australia was the whole problem of the foreign-ness of white bodies as *white*, their 'alien-ness' as well. Tropical medicine necessarily conceded this foreign-ness, constantly drawing attention to the settlement of Australia as a process of colonisation. The logic in the very idea of 'tropical' medicine always implied that white man did not really belong; or white man never quite naturally or easily fitted the space of the tropics. At best white man would have to continually monitor and regulate his conduct, his daily habits in order to live there. In other words, every time tropical medicine was called upon to demonstrate the capacity of white man to colonise the north, the north was reinscribed as *tropical*, not the place for white society. This was the ambivalence which characterised tropical medicine as a colonial knowledge. The question 'Is White Australia Possible?' was posed continually, one might say obsessively from 1901 and throughout the interwar years in the domains of tropical medicine, international hygiene, geography and public health. It seemed to function as a rhetorical device which invited repeated affirmation of white Australia – posed so that it may be answered over and over again in the affirmative. But there was always a difficulty at the heart of tropical medicine as a knowledge, as there is at the heart of any colonial enterprise: colonisation in Australia is exemplary here. The knowledge used to persuade white man that he *can* belong, was precisely the knowledge which historically set up the tropics as his other space, which made 'him', through this period at least, a foreign body.

7
Sex: Public Health, Social Hygiene and Eugenics

Administering the population with respect to sex in the first half of the twentieth century gave rise to all kinds of interventions in the governing of the self, as well in the securing and governing of nations. Several domains of regulation of sexual conduct and of population converged, each legacies of mid and late Victorian social questions, social science and social activism. And each problematised both the pathological and the reproductive implications of sex. Sexology – the science of sex – became expert and institutionalised. Linked but with a different genealogy was the very public feminist debate on venereal disease with its politicising of relations between men and women, between the forceful state and particular women, and its agenda for the self-governance of men in relation to their sexual desire. Interwar 'social hygiene' in Britain, can be understood as the outcome of one strand of sexology (concerned with birth control) and a direct descendent of nineteenth century feminist Contagious Diseases Acts protest. 'Social hygiene' in Britain was, as we have seen, 'racial hygiene' in Australia. Linked again, but with its own antecedents in nineteenth century biology, was eugenics. Because eugenics had the reproduction of the population firmly in sight, the position of women, the regulation of sex by individuals as well as by the state, and the nature of relations between men and women were issues which eugenic individuals and organisations constantly took up, hence the personal, intellectual and institutional cross over between feminism, sexology, eugenics and social hygiene. Finally, governments in the period – never separate from these other developments – had their own reasons for being interested in the regulation of sex, health and fitness. Anxiously concerned with degeneracy and falling birth-rates, many national governments were increasingly taking responsibility for reproduction as part of public health and welfare, in concert with the longstanding responsibility for the management of communicable diseases.

Separating the strands of these movements, organisations and intellectual histories of the regulation of sex, health and population is a teasing historical task, and certainly not my aim here.[1] I am interested, however, in exploring further the convergence of the fields, the problems and the logics of eugenics and public health.[2] If in the last chapter I explored this convergence with respect to international hygiene and migration issues, here I explore it with respect to the other biopolitical domains of sex, disease and reproduction. I am concerned with the increasingly calculated administration of contagious and reproductive sex: how eugenics and public health grafted onto the problematisation of sex, and thus grafted onto each other, as ways of comprehending, as well as acting upon, the population. In all, my interest is less to argue for the political and social reach of eugenics, as that has received considerable attention,[3] and rather more to explore the political and social reach of a eugenically inflected public health.

The management of the connection between sex, purity and reproduction ended up in many interwar contexts as a racism 'internal to the biopolitical state'.[4] The sense in which Nazi policies were born out of the human sciences, and were in part health policies concerned with strength of the population has been detailed.[5] Here I look at the dense clustering of the rationalities of the human sciences and the biomedical sciences which constituted a grafted eugenics and public health in Australia, but with reference also to British organisations and their imperial extensions. I am interested again in this chapter in the capacity of 'hygiene' to refer to many levels of governance at once: bodily and personal hygiene, domestic and urban hygiene, and as we have seen, imperial and international hygiene. 'Hygiene' in the interwar years is an exemplary site as well as explanatory tool, with which to explore governmentality, at once a technique of domination, and a technique of the (sexual, national, racial) self. Sex hygiene was a way of 'doing', as well as thinking, nationalism: it was also national and racial hygiene, at once an internalised ambition of citizen-subjects and an aspiration of government.[6]

Sociologists and historians of government constantly return to developments over the first half of the twentieth century to explore and explain the possible trajectories and the paradoxes of liberalism, and with good reason: these were the years of 'social government' and of welfare; of social democracy and compulsory sterilisation in Scandinavian nations and elsewhere; of commissions of enquiry into birth-rates, physical deterioration, the feeble-minded, lunacy and mental deficiency; of the gradual slide from the 'good despotic' rule which was always inside the logic of liberal government to the totalitarian capacity of the Nazi state to pronounce death as well as to administer life.[7] In short, this was the 'eugenic half-century' which saw the racialisation of many biopolitical states. Reading Foucault's *History of Sexuality*, Ann Laura Stoler both locates and extends his ideas about 'the intersection of sexuality, degeneracy and racism with the emergence of the biopolitical state'. This was 'a preoccupation with blood ... [which] haunted the administration of sexuality'.[8] Yet the blood lines Foucault was interested in were largely about preserving the pure past. Eugenic culture and its public health expressions problematised the sexual and racial connection between the present generation and its progeny – one, two, one hundred generations into the future. Eugenic culture expressed a desire for a future hygienic utopia, rather than a nostalgia for a pure past. This line between the present and the future, necessarily produced by sex, is the final line of hygiene I consider in the book.

Venereal disease: detention and education

In Chapter 3, on the sanatorium, I examined education into healthy and responsible citizenship as one of the objectives of segregation itself. This self-governance in matters of health proliferated in the interwar period around the pathological and reproductive questions of sex. What characterised public health and eugenics were elaborate campaigns of education, instruction and 'propaganda' to use the interwar term. Those well inside the social body (racially, mentally, physically) were positively bombarded with inculcation towards healthy habits and the civic education of desire for health and hygiene. I discuss this further below. But it is important to recognise the sustained medico-penal detention of certain sub-populations as a way of qualifying the common argument that there was a modern shift toward a 'governmental' new public health in the twentieth century.[9] The ongoing place of segregative medico-penal governance can be illustrated especially through the problem of venereal disease management – the crux of 'social hygiene'. In concert with increasingly refined classifications, the removal and compulsory treatment of the medically and morally dangerous was sustained. Indeed, eugenic mentalities fostered and facilitated the classificatory and segregative tendencies which, as we have seen through the book, had long belonged squarely to the field of public health. Thus, far from analysing such public health detention as a residual legacy of Victorian CD Acts or of 'old' quarantining technologies, it is more accurate to see such trends as consistent with the fresh rationales which eugenics brought to public health classification and segregation.[10] By the same token, it shows the way in which eugenic ideas and practices absorbed the classificatory and segregative ambitions of public health.

Throughout the book I have examined different legal and cultural genealogies of public health detention. The emergency quarantine enacted in the early nineteenth-century European cholera epidemics were the legislative precedents and the carceral models for later management of acutely infectious diseases in Australia (and elsewhere): for example, smallpox in 1881 and 1913, plague in 1900, influenza in 1919. I have also examined the hybrid nature of the tuberculosis sanatorium as a disciplinary space: part health-resort, part quarantine station, part workhouse/asylum, part farm colony. Asylums and Lunacy Acts also offered rationales for and means of segregation.[11] Lock hospitals for prostitutes with venereal diseases served as a further model for the public health detention of people with chronic diseases, if a controversial one. On the one hand the anti-CD Act agitation in Britain haunted twentieth-century public health in all kinds of ways. But on the other hand, the lock hospital was sustained as a model, as well (in some locations) as an actual practice for the removal of the morally as well as the medically dangerous. This was especially so in the first half of the twentieth century when categories and classifications of dangerousness and unfitness were authorised and implemented far more rigidly and finely than in the nineteenth century, and when exigencies of wartime made such classification and spatial segregation seem even more socially necessary, for many.

Attempts to manage venereal disease in the nineteenth century gave rise to many of the problems of liberal governance that I have discussed. Along with compulsory vaccination, the compulsory detention, examination and treatment of certain women in certain places under CD Acts throughout the Empire provoked debate over personal sovereignty, the legitimacy of public health detention in lock hospitals, and the controversial suspension of habeas corpus. There was also specifically feminist protest over the surveillance of women, not men in the circulation of the disease and in the regulation of sex. The strength of this opposition meant that compulsory detention was deemed an impossibility thereafter in Britain itself, but elsewhere in the Empire and the Commonwealth, this version of public health detention was sustained well into the twentieth century. As Philippa Levine has shown, the methods, the problem and the heatedness of any CD Act agitation varied across imperial locations, and in quite a few cases, the colonial acts were sustained well past their domestic repeal.[12] In fact, to take the British Empire as a whole, lock hospitals were in and out of use from the early nineteenth century – in some Indian provinces for example – and until after the Second World War – in some Australian states.[13]

Unlike the British instance where a strong combination of Victorian liberalism and feminism made public health detention for venereal disease virtually impossible to contemplate in the twentieth century, in Australia, medico-penal systems were either newly implemented in the early twentieth century, or were sustained from their original implementation in the 1860s. In Queensland, the Contagious Diseases Act of 1868 was modelled on the series of British Acts, but was significantly different. If the British Acts were limited to the compulsory detention, examination and treatment of women within certain garrison towns, the Queensland Act could be applied, in theory, to anyone, anywhere. It was replaced by the Queensland Health Act of 1911 which, while omitting the objectionable reference to contagious diseases and their problematic history of management, nonetheless maintained the lock hospital system. Women detained under the 1911 Act who were also categorised as 'common prostitutes' were placed in the Female Venereal Disease Isolation Hospital. In 1913 the hospital was moved from the medical site of the Brisbane General Hospital to the penal site of the city's main prison complex. There, women with venereal diseases were compulsorily held in a ward behind the central building for male prisoners.[14] On the one hand this spatial conflation of medical and penal institutions and practices was longstanding and not uncommon, as we have seen throughout the book.[15] But on the other, it was a significantly *new* legislative and geographic/architectural expression of this conflation.

A further example from the period concerns Aboriginal people, an instance of the kind of racial *cordon sanitaire* discussed in Chapter 4. Driven by the desire to minimise venereal disease amongst Indigenous communities, officials in the Western Australian government sought ways and means to effectively isolate afflicted Aboriginal people. One suggestion made in 1906 was to 'declare a reserve and muster in it all the natives and have a thorough examination made of them all'.[16] This would be a two-acre block with a ten-foot fence surrounding it, and with a dividing fence to separate men and women. The idea was rejected, however, in favour of the use of islands. What is interesting here for my purposes are the legislative moves made to legalise this particular compulsory detention. The islands themselves were imagined as lock hospitals – that is, as public health/medical spaces in the carceral tradition. But they were *made* carceral not through public health legislation or even criminal/penal legislation, but through Aboriginal protection legislation. A 1911 Amendment Act was passed specifically to make the islands, and equivalent lock hospitals on the mainland, 'reserves', thereby giving doctors and the state the right to detain Aboriginal people in them.[17]

The New South Wales Prisoners' Detention Act of 1908 is a final example of the ongoing carceral nature of public health spaces. Beginning its passage as the Contagious Diseases Bill, its title was later altered, legislators again eschewing direct reference to, and replication of, the CD Acts. But the change to 'Prisoners' Detention' reveals how closely aligned the medical and the penal systems were. In its final version, this was a disguised piece of venereal disease legislation. Designed to work in concert with the Police Offences Act, which rendered soliciting punishable by imprisonment, it allowed any prisoner to be inspected for venereal disease. If infected they could be detained in prison until the disease was cured, that is beyond the period of detention of their original sentence.[18] In this version, prisons themselves *became* the lock hospitals, and many people were imprisoned in order to be rendered into 'patients'. The Act empowered the Governor to proclaim 'any hospital, or any part of a hospital or of a public gaol, prison, or house of correction, or of a place of detention to be a lock hospital'.[19] After extensive discussion of this legislation and venereal disease management at a Select Committee Inquiry in 1915 chaired by Member of Parliament and President of the local Eugenics Society, Dr Richard Arthur,[20] a new Venereal Disease Act was passed in 1918. Partly because of the exigencies and the culture of extraordinary measures produced by the War, the 1918 VD Act was a heavy-handed piece of legislation. It rendered notification compulsory, required that a person with any form of venereal disease must consult a doctor or hospital with financial penalty or three months imprisonment. Further, a person infected with venereal disease 'must not marry', or face a penalty of £500 or five years imprisonment or both, and must not knowingly infect another person.[21]

These carceral practices and spaces were newly authorised in both world wars. In the Great War, for example, returning soldiers with venereal disease were generally held in camps until pronounced cured.[22]

Judith Smart has demonstrated how venereal disease policy in Australia 'ran parallel to the changing assessment of the defence needs of the colonies and nation and, with those needs, developing ideas about the specifically gendered duties and requirements of citizenship'.[23] In 1942 there was a new regulation under the wartime National Emergency Act which permitted the detention and compulsory examination of any venereal disease suspect throughout Australia. It was in force especially strongly in Queensland. In Australia such places and practices of detention as a way of managing venereal disease continued until 1946, with the repeal of this Act. From that period penicillin for syphilis and sulphonamides for gonorrhoea quickly and radically altered the longstanding spatial and carceral organisation of venereal disease prevention, although the use of the logic of 'dangerousness', it has been argued, persisted into the early management of HIV/AIDS and indeed into the twenty-first century logic for detaining asylum-seekers.[24]

Rather than indicating a phasing out of 'quarantine' technologies over time, these instances show how carceral spaces of public health were being reinvented, given fresh expression in the first half of the twentieth century. Not only were there new institutions for categories like the feeble-minded, for epileptics, as well as the infectious disease hospitals, the leper colonies and the sanatoria that I have discussed, there were new sites for the holding and treating of certain populations with venereal disease, mainly women defined as common prostitutes but also men as returned soldiers.

In Britain, the nineteenth century controversy over the CD Acts meant that compulsory examination and detention was highly unlikely to be tried again domestically, and there was, as we have seen in the case of leprosy, a general trend away from enforcing other kinds of public health detention. 'A Threat of Compulsion!' was the shocked and exclamatory leader of a 1937 issue of the British Social Hygiene Council's *Health and Empire*. It was imperative for the Council that voluntariness be its core philosophy. Taking aim at the Queensland government, amongst others, the editor wrote: 'Practically all other countries have thought it necessary to have compulsory powers, and even our own Dominions have taken a different line from the mother country on this point'. The article stressed that 'a voluntary scheme is in line with British tradition, which puts great stress on personal liberty'.[25] If other countries compelled the inspection and detention of prostitutes, there was an investment in this public health context, as in Sir Leonard Rogers' leprosy plans, in British 'freedom'. British rule, liberal rule ideally sought consent through education and opportunity.

Enlightened sex and health education, not detention and inspection, was the mission of the Social Hygiene Council. It defined social hygiene as measures 'associated with matters of sex in its various aspects, physiological, psychological and morbid, especially as these affect social and family life'.[26] Importantly for my purposes, the Council comprehended venereal disease and its management squarely as an imperial problem and responsibility. Colonial tendencies toward compulsion, as well as the disease itself, needed to be countered by 'the enlightenment of the population by specially arranged educational and propaganda methods … the development of a lively health conscience'.[27] The ethics of conduct had an explicitly imperial reach. The Council organised the Imperial Social Hygiene Congresses regularly through the interwar years, thereby securing 'Imperial co-operation in public health'.[28] New technologies were developed alongside the idea of 'public instruction' which governments inherited from nineteenth century philanthropic and sanitary associations. In particular the educational film on venereal disease prevention and sex conduct was the highly favoured tool of instruction. The Social Hygiene Council sent out films to all the dominions and colonies, usually utilising the equivalent local association. The 'Empire Educational Projector' was advertised in *Health and Empire* as 'a new power for propaganda work'.[29] This education into healthy conduct was understood as an imperial responsibility, part of the liberal program of educating into the capacity for citizenship: 'British policy gradually confers the responsibility of citizenship on the various races for whom it acts as trustee. It becomes therefore of primary importance that citizens of British and of other races should have some knowledge of the social problems and of the methods of handling them that have proved effective'.[30]

In Australia, education as venereal disease prevention also concerned eugenicists, feminists and nationalists. While compulsory detention as a method of prevention was problematic for some, the coercive and persuasive systems coexisted far more comfortably in Australia than would have been possible in Britain. The Racial Hygiene Association initially referred to itself as the Race Improvement Society. It was formed for 'the teaching of sex hygiene … to work for racial health improvement, which includes sex education and the eradication and prevention of Venereal Disease'.[31] Like the Social Hygiene Council (and indeed many other similar interwar associations),[32] the venereal disease issue was for this Association a gateway to the other great concerns of the period, precisely because it was about reproduction and hygiene. All kinds of social problems exercised the members of the

Racial Hygiene Association: the coercive treatment of Aboriginal people, the age of consent, the programmes of assimilation, sterilisation and mental hygiene (they moved in favour of compulsory sterilisation), birth control and as we have seen, immigration and health screening. The Association secured many of the Social Hygiene Council's anti-VD films, showing them mainly to inner-city audiences. They hosted a weekly talk on the theosophist radio station 2GB, which was to have a long history of broadcasting social hygiene and sexological issues. In all, at the same time as the 'unfit' and the morally and medically dangerous were being sought, diagnosed and managed spatially in Australia, the fitness of the 'fit' was constantly being urged as a personal responsibility.

Reproduction and responsibility

By the twentieth century, government interest in regulating sexual conduct incorporated not only the longstanding issue of sex as a potentially pathological act – venereal diseases – but sex as a reproductive act – the future population. Reproduction of the population was crucial for the biopolitical state. The idea of medical police in the eighteenth century held clear pro-natalist ambitions in which the promotion of marriage, the state supervision of midwifery, and regulation of early infant welfare practices were priorities of government.[33] Indeed this plan of 'medical police' was all but realised in the pro-natalist and later the eugenic twentieth century, when reproduction was brought into the fold of government and public health as it had never been before. For many nations, the early welfare state was a maternalist state, a 'pro-natalist' state. Across the western world, the precise programme of eugenics and its biologically based theories of promoting good breeding and preventing bad breeding in human populations merged into that turn-of-the-century moment when governments and experts were acutely concerned, in the first instance, with simple numerical assessment of population. Declining birth-rates of nationally bounded populations – often figured as 'racial suicide' – was the early governmental issue.[34] In England class differentiated birth rate was related to the *quality* of the national health as a whole, and this deeply informed new health and welfare policy. Reproduction was part of the health of the nation as well as the Empire.[35]

In Australia a marked decline in women's reproductivity was taken very seriously by State and Commonwealth governments who projected their statistics both temporally and geographically onto (what was seen as) the hopelessly underpopulated 'empty' new Australian space. Government statisticians and health experts gathered information on birth-control methods and usage, on abortion, on infant and maternal mortality rates, on venereal diseases and on techniques and places of confinement.[36] In the interwar years, reproduction, childbirth and infant welfare became bureaucratised and increasingly regulated within departments of health. For example, the New South Wales Labor Government that established a very early Motherhood Endowment Scheme, created in 1920 a new Ministerial Portfolio of Public Health and Motherhood. Race-specific welfare policies were developed to support and encourage reproduction of future citizens.[37] Conduct in relation to motherhood became a highly regulated industry as health departments sponsored centres and experts in maternal and infant welfare, as feminist groups created new women's hospitals, and as an industry in mothering advice literature was advanced.[38]

Women became citizens in this period in many ways as mothers, and governments rapidly assumed financial, regulatory and institutional responsibility for the promotion and conditions of childbirth as well as reproductive sex. This was as true of socialist and liberal governments as of interwar totalitarian governments.[39] Women's reproductive selves were thus increasingly the objects and subjects of biopolitical governance. National or racial hygiene *became* the responsibility of individual citizens, insofar as they could exercise choices over their reproductive conduct and their bodily fitness. Healthy habits and conduct were made desirable for the dutiful citizen, a simultaneous desire for self-fulfilment and of personal responsibility for the nation/race. As the editor of *Health and Empire* put it, the aim of social hygiene and public health was to make 'healthy living and healthy thinking ... recognized as an essential equipment of the citizen'.[40] Thus the period is characterised by an extraordinary amount of instruction into, and cultivation of, reproductively responsible and fit selves. The British Social Hygiene Council took its 'propaganda work' extremely seriously, both its mission to the British population and especially its mission to educate imperially. The Propaganda Committee suggested 12 points that should be stressed in Social Hygiene Propaganda work. One of these was to 'promote an attitude of mind towards questions of Sex that attaches individual responsibility to the exercise of the racial instinct'.[41] The Social Hygiene Council aimed to 'emphasise the responsibility of the community and the individual for preserving or improving, by educative and social measures, the quality of future generations'.[42] A company creating Health Illustrations advertised with the

Eugenics Education Society, producing images and posters which impressed on the public conscience not only 'the Delight and power of perfect health' but 'the fact that each person's health is mainly in his own hands'.[43] For the great mass of people within the Australian or British civic body and throughout the Empire and Commonwealth, this was the era of education into desire for fitness and healthiness, the era of modernised health propaganda, posters, Health Week, public instruction, the National Health Campaign, the era of films, lecturettes and radio talks on health and hygiene matters.

One site of intervention was marriage itself – the major institution for the regulation of reproduction. As one New South Wales birth control advocate rallied in 1932: 'To reach and influence all the dysgenic influences which have invaded the body politic of health, we must attack one of the chief "danger zones" without further delay – the zone of marriage'.[44] The Racial Hygiene Association of New South Wales established clinics in the 1930s for precisely this kind of marital advice.[45] It both counselled and tested couples, as well as created and promoted public education on the responsibility to seek and declare knowledge of one's own or one's family's physical and mental history. Undertaking this risk management procedure, those intending marriage could complete a mental test for personal or familial traces of epilepsy, perversions, alcoholism or inherited tendencies and a physical test for syphilis, gonorrhoea, tuberculosis or again, 'inherited diseases and tendencies' (see Figures 7.1 and 7.2). Marion Paddington, a birth control advocate who set up a rival clinic to the Racial Hygiene Association, wrote that 'the task at present is to safeguard the future of those entering marriage today. This may be done by securing blood tests from both ... Such deleterious strains as mental deficiency, congenital venereal disease, tuberculosis, epilepsy and alcoholism should not be allowed to persist'.[46]

There were attempts to make this kind of health certification compulsory, although in Australia this never eventuated. Richard Arthur, sometime Minister of Health and President of the New South Wales Eugenics Society, argued for its compulsion from the time he chaired the Venereal Disease Select Committee in 1915 throughout the interwar years. If 'clean bills of health' (interestingly using the old port term for vessel clearance) could not be compelled, he argued to the Health Week Conference in 1931, 'we can spread the knowledge that it is extremely desirable ... We must think of the great risk not only to the one partner, but also to their possible offspring'.[47] In 1926 a 'Qualification for Marriage Act' was presented to the Eugenics

Figures 7.1 and 7.2 Pre-marital health screening: the eugenic *cordons sanitaires* between the current and future generations
Source: The Racial Hygiene Association of NSW *Annual Report*, 1938–39. Courtesy, the Mitchell Library, State Library of New South Wales

176 *Imperial Hygiene*

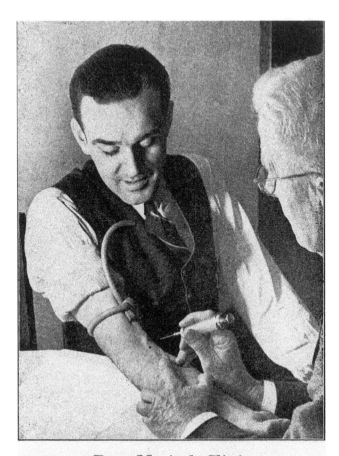

Pre - Marital Clinic
PHYSICAL TESTS

Support the Racial Hygiene Association of N.S.W. in its crusade for happy marriages and healthy children, by means of the exchange of health certificates. Avail yourselves of this opportunity to gain knowledge of yourselves, mentally and physically.

General Physical Examinations include :
- Blood Tests.
- Chest X-Ray.
- Blood Pressure.
- Diabetes.
- Bodily Defects.
    - Etc.
- Inherited Diseases and Tendencies.

Figures 7.2

Education Society in London from a Melbourne-based member. It turned out to be no such thing, but rather one eugenicist's expression of what *should* be enacted in Australian parliaments, as elsewhere. 'No persons may contract marriage ... unless ... a sealed Certificate is issued ... that such parties are fit and proper persons to marry'. Under his scheme, 30 days before intended marriage, the couple would be required to give details of personal and family history. No certificate of fitness would be granted 'if either of the parties is suffering from any infectious and contagious disease, or is deformed, or mentally deficient, or degenerate' or if any child would likely become 'a burden on the State'. Importantly, these categories of unfitness were identical to and probably drawn from, the health restrictions placed on entrants under quarantine and immigration powers. As with many eugenic arguments, the link was drawn here between such a requirement of individuals and the viability and healthiness of the citizenry, which in his opinion needed to be more exclusive. Radical democratisation, he argued had meant that 'increasing voting power is placed in the hands of those most unfitted to exercise such power'.[48] If one could not wind back the unwise extension of the franchise, one should shape the fitness of the future citizenry. For him, the eugenically healthy citizen was the only legitimate reproductive citizen.

Like compulsory certificates of health for marriage, some argued for compulsory sterilisation of the feeble-minded. Between the respective Labor Party and Roman Catholic opposition, as well as ongoing debate about the compulsion of vaccination, compulsory sterilisation was effectively avoided in Britain and Australia, but not, as is well known, in Nazi Germany, in the Scandinavian nations, in provinces of Canada, and several US states.[49] Rather, it was the cultivation of individual 'racial' *responsibility* around sex which characterised the eugenic moment in these national contexts, alongside the widespread use of segregation of those deemed outside the civic body. 'Voluntary Sterilization for Human Betterment', the Vice-President of the Victorian Eugenics Society urged in a 1938 lecture.[50] This was a liberal version of governance through freedom, of encouraging and educating into the desire and need for responsible sterilisation (or vaccination or health certificates). These were not only civic duties, but actively constituted an individual as a good citizen. Paradoxically, one could even be educated into the civic duty of accepting compulsion. In 1947, for example, the Eugenics Society in Victoria again broached compulsory health checks before marriage – X-ray for tuberculosis, blood test and examination for, and family history of, inherited diseases. It was suggested that 'such examination should be free and public opinion should be educated to the point where it would be prepared to accept it as compulsory'.[51]

In addition to the proliferation of programmes, institutions, and literature in maternal and infant welfare in the period, government sought to shape men's conduct in relation to reproductive sex. Sex education for boys was a feature of the period, and was constantly argued over in terms of nature, style and content. Government departments of Health and of Education were increasingly co-operating and crossing over, not only with respect to physical issues – diagnostic procedures for diphtheria, dental hygiene checks, the supply of school milk – but also with respect to moral and sexual instruction.[52] This was instruction in civic responsibility, gathered under the idea of hygiene. 'Education of the community in hygiene', one commentator put it at a conference held in Health Week, 1931. 'Educationists are charged with the task of implanting the threefold ideal in the child mind – individual responsibility, community consciousness, and racial welfare'.[53]

Men were also enjoined to take real responsibility for themselves as reproductive, and therefore raced and national, beings. Citizens of New South Wales were asked by the Health Department to 'Sow the Seeds of Good Health', a poster campaign of the interwar years (see Figure 7.3). Here health and the reproduction of healthiness converged: one needed to be healthy in order to responsibly and dutifully pass on healthiness to the future generation. The Realist style of the health poster, now so strongly linked to Soviet propaganda as well as Nazi hyper-masculinity,[54] indicates how the form and the content belong as much to the interwar period generally, as to the totalitarian regimes specifically. That is, the connections between public health, national strength, and reproductivity made so powerfully in this poster held as much currency in interwar New South Wales as they did in Britain or Nazi Germany, if with different effects. There were several messages of responsibility carried in the image and the text: the responsibility to think of sex as reproductive; within that idea, the responsibility to understand sex as potentially reproducing infections, inheritable traits, tendencies, and undesirable qualities in the next generation; and therefore the responsibility to know and secure one's own health. Responsible sex reproduced and enhanced the social, national, and racial health and vitality, and as this image suggests, strength and virility. In this way, good citizens were not just cautioned negatively to ensure that they did not pass on taints, but enjoined positively to pass on the seeds of good health: 'laying upon the conscience of the people

Figure 7.3 Nationalism, virility and responsibility: Sow the Seeds of Good Health
Source: New South Wales Department of Health poster, c. 1930. Courtesy, the Powerhouse Museum, Sydney.

180 *Imperial Hygiene*

their responsibility to bestow upon their children the endowment of their own health in every part of their being'.[55]

Particularly important here, is the way in which many governments, alongside organisations like the Eugenics Education Society, the Racial Hygiene Association of NSW, the British Social Hygiene Council and others, saw 'education' and 'habit' as both ends and means. These were the mechanics, the practices by which individuals participated in the biopolitical shaping of population: in these ways a civic identity implying both responsibility and health was both asked of people, and created. In this period of intense nationalism, one's own intimate sexual choices and actions were understood to be always significant for the nation/race. Civics, public health and eugenics went together.

The eugenic *cordon sanitaire*: contagion and the future population

By thinking about public health as part of an interwar and an imperial eugenic culture, it becomes possible to understand eugenics itself as a kind of preventive health. This is so simply at the level of substantive topics of population management, which crossed over the domains of public health and eugenics: the interlinking population issues of reproductive sex, venereal disease, as well as tuberculosis and migration regulation, which I have intermittently discussed. But the grafting of eugenics onto public health is also important to analyse at the level of problematisation and technique: the problematisation of connection, contagion and contact, and the complementary hygienic technique of the *cordon sanitaire*. In that the reproduction of the population was so important in this period, and in that all kinds of social and bodily qualities were being refigured as inheritable, that is, as contagious between generations, it is possible to understand the many regulatory techniques in play as eugenic *cordons sanitaires*.

The eugenic *cordons sanitaires* took several forms. They incorporated the older mechanism of lunacy incarceration. But less recognised is the extent to which they borrowed from, indeed became part of, the spatial mechanisms of communicable disease control. That is, they were plainly segregative. Some segregative measures were enacted and implemented as social policy, others were the dreams of social planning utopians, dreams of the perfectibility of man, of nation or of race. But I am interested here in drawing the connection between longstanding spatial public health measures, and the way in which they merged with eugenic hopes of segregation with the aim of preventing sex and reproduction between the fit and the unfit, however classified. In crudest form, the unfit were isolated from the fit so as to separate the lines of descent, ideally to block any line of unfit descent. This manifested as institutions and spaces like the epileptic colony, and asylums for the mentally unfit.[56] To some extent the leper colonies also came to be considered under this rationale, because of the still strong comprehension of leprosy as hereditary. What had been implemented under other preventive logics, were drawn into the eugenic logic of segregation: the prevention of reproduction between the fit and the unfit. Moreover, reproduction *between* the isolated unfit was likewise prevented spatially. The supervision of the segregation of the sexes which had long been typical of penal and medical institutions of isolation, was given added significance by the eugenic concern for the future population.

Over the first half of the twentieth century, segregation and sterilisation came to be discussed as alternative methods for the prevention of the spread of physical and mental taints into the future generation. 'There are two possible ways to prevent the marriage of the feebleminded, namely segregation and sterilisation', it was put succinctly by a speaker at the 1931 Health Week conference in Sydney.[57] Indeed sterilisation was often understood to be an inexpensive, but also a more humane alternative to institutional isolation.[58] At other times, sterilisation and segregation were perceived and pursued as twin measures.[59] In either case, sterilisation was another form of the eugenic *cordon sanitaire*. It severed the link between the unfit individual and their now impossible offspring. That particular problematic descent line was brought to an end with a eugenically satisfying cut.

Mitchell Dean and others have written about the sovereign power of death that has historically accompanied the biopolitical administration of life in certain circumstances. In the twentieth century, he writes, killing itself 'is re-posed at the level of entire populations ... This power to disallow life is perhaps best encapsulated in the injunctions of the eugenic project'.[60] Of course qualitatively and ethically entirely different to the experience of Jews and others under National Socialism to which Dean largely refers, the Australian and British eugenic *cordons sanitaires* of segregation and sterilisation were both nonetheless ways of systematically disallowing life in the interests, as it were, of a particular vision of a future population. As we have seen – and of course this is very important – compulsory sterilisation was never enforced in Australia or in Britain, but this was not for lack of effort on the part of many health bureaucrats, eugenic advocates and social/racial hygiene organisations.[61] Nor did the failure to render sterilisation of the mentally deficient compulsory mean, of course, that it did not take place within all kinds of institutions, legalised with the consent of doctors, magistrates and parents or state guardians. Voluntary sterilisation was also undertaken as a form of birth control, along with 'barrier methods' another kind of *cordons sanitaires* with the future.[62] It is the segregating logic of the techniques of both isolation and sterilisation which is significant here, for in both discussion and implementation the logic belonged to the pre-existing field of health and hygiene.

In Australia, experts looked not only to internal segregation or sterilisation, but also to immigration/quarantine lines to secure the mental and physical hygiene of the population. In this way, immigration restriction laws and regulations were not only imagined as barriers against the introduction of infectious diseases, but increasingly strongly, as discussed in the last chapter, as eugenic *cordons sanitaires*. The immigration/quarantine regulations were also called up as ways to (re)produce and retain racial homogeneity within the population, and to prevent interbreeding between the internal white race and external other races. Simply, migration lines were one mode of racial segregation, and not only in the Australian context. The Eugenics Education Society in its pamphlet on responsibility to the future generation, *Those Who Come After*, aimed 'to prevent the immigration of undesirable aliens to this country and so prevent the debasement of the race by intermarriage'.[63] In H.L. Wilkinson's book on global population issues and white Australia, the connections between these different lines of hygiene, these related *cordons sanitaires*, structured his chapter outline. Logically for him, as for many others in the period, a chapter on 'Immigration Restriction Laws' throughout the world was followed with a related chapter on 'Interbreeding and Segregation of Races'.[64] For him, immigration restriction laws *were* the major 'barriers to racial interbreeding' on a global scale, barriers which the 'colour bar' in the United States and in South Africa emulated.[65] In such ways, immigration borders were quarantine lines, were colour bars, were eugenic *cordons sanitaires*.

In many instances, prior segregative practices on the grounds of health and infectious disease management, as well as prior segregations on the grounds of race, assumed eugenic rationales. Conversely, strictly eugenic policies or plans assumed health measures as well as the language of contamination and segregation. One text, *Future Generations: Women the Future Ruler of this Earth*, put it this way:

> A leper – diseased – is isolated from his fellow beings and not allowed to perpetuate. A social leper (syphilitic, epileptic, etc.) may

roam at will and leave progeny where and when he wished to the detriment of future generations ... If the man and woman are honest, true, sound mentally and bodily their progeny is required by future generations, but if they are diseased mentally physically or morally <u>FUTURE GENERATIONS DEMAND THAT THEY SHOULD NOT BE ALLOWED TO PERPETUATE.</u>[66]

The author of this tract argued that the one ethical obligation of all humans is to future generations. If one related unethically to those generations by procreating with taint, one should be punished and rendered incapable of reproduction: 'If after punishment man or woman still showed signs of disease they should be treated as diseased lepers and isolated – removed to a portion of the earth reserved for such diseases (lepers) persons [sic]'.[67]

This was an extreme and idiosyncratic text. Yet it accurately suggests the common eugenic (indeed public health) centrality of women as reproducers, or of Woman, 'a creature to whom the race is more than the individual', as another writer put it.[68] It also accurately indicates the driving eugenic concern for unborn generations. An ethical responsibility to the future generation *did* in many countries outweigh the civil liberties of individuals in the present. 'From Generation to Generation' was the title of but one educational film, shown by the Eugenics Education Society in Victoria.[69] And the inside cover the of the British Society's publications *Those Who Come After* reads: 'Our duty to posterity is at least as great as our duty to our neighbour'.[70] It was put in a detailed way by Dr Richard Arthur in a parliamentary speech in 1921: 'we should concentrate our attention on any Proposal which will have the effect of benefiting the coming generation ... We owe to the child, if possible, that it should have good parents'. He then got specific: 'We will guard against persons who are suffering from syphilis, from tuberculosis, or who are mentally defective in any way will hand on some vicious or defective trait to their children – from being able to reproduce their like. There is no doubt that the segregation of such individuals will be the strong aim of the future community'.[71]

The eugenic mentalité created contagions out of many human qualities, not just transmissible diseases like syphilis or tuberculosis. All kinds of previously social and moral attributes were reified into physical and later genetic attributes which could be passed on across space or across time. On the one hand this was driven by longstanding cultural usage of hygiene metaphors, or of folk theories of contamination.[72] But it was not just that in this period. Epilepsy, feeble-mindedness, homosexuality, criminality, prostitution, alcoholism, were all rendered into pathologies considered transferable between generations through reproduction. They were 'caught' by one generation from another. This literal not just metaphorical pathologisation of certain attributes into transmissible phenomena is yet another reason for the convergence of public health and eugenics, under 'hygiene'. Hence the need, in 'marital hygiene', for screening in all these mental and physical conditions. In controlling 'mental defectives ... either by confining, or in certain cases sterilisation [we] prevent them from at least handing on their defects. Many defectives become criminals, unemployables, and prostitutes who spread V.D.'[73]

The 'reproduction' of these pathologies was understood in ways which confused the mechanisms of heredity with those of infection. Conditions like syphilis or gonorrhoea were understood as 'inheritable', when what was meant was the transmission of the microbe from the mother to the child at birth. Yet the connection between heredity and infection was not just a naïve mistake in the period, but rather a longstanding interest in and effort to understand the relations between what is now called 'horizontal transmission' (between people in the present) and 'vertical transmission' (between generations): the relations between heredity and infection.[74] That so many qualities of human being were pathologised and newly comprehended as transmissible, whether genetically, natally, microbially or even psychologically, illuminates the oftentimes perfect fit between eugenics and public health. When the reproduction of a taint was comprehended as a process of contagion this invited, logically, public health measures in response. Segregation, sterilisation, migration regulation, education in the management of one's reproduction, linked as *cordons sanitaires* between the current and the future generation, were useful and seemed entirely appropriate for the new eugenic culture of managing the health of populations.

Conclusion

In the eugenic half-century there was an increasingly refined classification of people *out of* the civic body: the leper, the mentally deficient, on occasion the syphilitic and the epileptic were all put under different rule, often, even usually, through spatial segregation. But the period is also characterised by the flourishing of education, instruction and consent-seeking propaganda around health: governance through the powers of freedom and of instilling responsibility.⁷⁵ Modes of rule differed radically according to one's position within or without the civic body. The problematisation of sex in the eugenic half-century – as contagion and as reproduction – illustrates clearly these modes of governance working side-by-side.

If public health was concerned primarily with the health of the present population, eugenics concerned itself with the health of the future population. The Eugenic Education Society differentiated itself from the project of 'sex hygiene' expressly because it looked to the future, rather than the present. But they were linked because 'the mechanism of sex ... connects one generation with another'.⁷⁶ And if eugenics problematised the biological connection between the present and the future, its mode of action to shape that connection can be understood as the erection of *cordons sanitaires* between its present and that projected time, in particular a break in the reproduction of race contagions. Breaking the reproductive connection between now and the future was what characterised many eugenic technologies of intervention. But segregation and the use of the *cordon sanitaire* in the interests of population health had long been the business and the technique of public health. It is this convergence of both objectives and means, which facilitated the relatively easy mapping of eugenic ideas onto pre-existing public health technologies. This is partly why distinguishing between eugenics and health in the period is so difficult, and perhaps a misplaced ambition. Like public health, eugenics wanted to stop the connections and circulations of contagions through segregation, sterilisation, migration lines, isolation or education. Each of these was in one way or another a eugenic *cordon sanitaire* between the present and future population.

Conclusion

Over the modern period, public health was oftentimes imagined and implemented by forceful states, through policing, detention and compulsion, through the identification of the dangerous and the suspension of habeas corpus. But another genealogy of public health was that of 'governmentality', in which practices of the self and the objectives of expert agencies came increasingly into alignment. In many ways it is more useful to understand these changes over time not as a broad movement from the former to the latter, from 'quarantine' to the 'new public health', from sovereign powers to (self) governance of conduct, or even from dangerousness to risk. Rather, I suggest imagining the manifestation of these modes of power in three kinds of spatial practices, which moved forward in parallel over the modern period, but with different weightings and respective intensities at different moments. First, there were places of coerced segregation and institutionalisation or exile-enclosure (here examined in smallpox quarantine, venereal disease lock hospitals and the island leper colonies). Second, there were segregative practices in which people's consent was sought by experts and governmental authorities and where the purpose of segregation was not simply removal, but reform (here examined through the sanatorium for consumptives). And third, there was education into a desire for health and fitness and into certain kinds of conduct amongst the general population and in 'public' space (here examined through civic and sex education). There are elements of each of these forms of governance over the whole period, but they moved forward in changing relations to each other, and importantly, with changing significance for, and applications to, different populations.

In part it is the colonial context of a white 'settler society' which matters here, and which clarifies the distinctions between populations subject to each of these forms of rule. In the first instance, the racial, nationalist and eugenic politics of the Australian context exposes ongoing uses, and indeed innovative measures of public health detention well into the twentieth century. I have traced here *new* powers of enclosure, segregation, and classification of the medically dangerous. Especially in the interwar period, the liberal interests of certain individuals were routinely subordinated to communal interests (the race, the nation, the Empire). There is, then, a rather more continuous and complicated genealogy of confining the microbially dangerous – explored in contemporary instances with respect to HIV/AIDS, multi-drug resistant tuberculosis and most recently SARS[1] – than is often recognised. But I have also given analytic attention to the specific populations for whom, and ways and moments in which, isolation and containment measures were deemed illiberal and abandoned, or in which they came to be entered into 'voluntarily', as subjects of freedom. For such subjects health and hygiene were sometimes instilled in institutions and enclosed spaces, and sometimes in the public spaces of health education and propaganda. This 'public' health was in fact intensely private, aiming to shape and form personal habits and conduct. Public and private cleanliness were linked, and were both the business of government.

'Cleanliness' is the title of an undated typescript by J.H.L. Cumpston. Evidently composed in the 1920s as a talk to a meeting of interested lay people, Cumpston divided his paper into five sections: 'Personal Cleanliness', 'Domestic Cleanliness', 'Communal Cleanliness', 'National Cleanliness', and 'Imperial Cleanliness'. In its capacity to invoke the interconnected governance of the individual subject with the population and the nation, as well as colonial rule, Cumpston's musings on cleanliness and hygiene capture the historical and sociological significance of public health for the modern Western world, a significance I have sought to explain and demonstrate through the book. He opened his speech with a schema, conventional enough for the time, about the relationship between 'man' and dirt: how in a state of nature, man lived in an open air environment 'almost free from bacteria',[2] but that in the artificial urban clustering of civilisation there is an accumulation of dirt of various kinds. In order to survive such artificial conditions, constant and vigilant cleanliness is required, at all the levels specified, 'it is cleanliness on a large scale which is necessary.'[3] Cumpston understood this vigilant cleaning as productive, as improving and strengthening the capacities of the population, or 'the race' in his terms: 'the physique, stamina, and mental power of the race, will be very greatly improved if people will make up their minds to be clean themselves and to live in clean surroundings'.[4] For Cumpston, and so many of his contemporaries, attending to cleanliness – hygiene – always had a reach far greater than oneself. As related throughout the book, 'National Cleanliness' was an obsession in this period of intense, and squarely race-based Australian nationalism. He nominated two specific practices as crucial: quarantine, which he described as 'the keeping of our continent free from certain deadly diseases at present unknown amongst us'; and immigration restriction, which he detailed as 'the strict prohibition against the entrance into our country of certain races of aliens whose uncleanly customs and absolute lack of sanitary conscience form a standing menace to the health of any community'.[5] Cumpston finally turned to 'imperial cleanliness' and articulated a view of the British Empire, especially the 'tropical portions' of the Empire, in which public health and sanitation were to 'take precedence over everything else'. This form of 'colonisation', in his words, was 'sane imperial development'.[6]

For Cumpston and many other theorists and practitioners of public health in the period, hygiene was formative of subjects and formative of nations. Indeed his understanding of 'cleanliness' is a perfect articulation of the connections implied in governmentality, the simultaneous imperatives of individual conduct and political administration. Despite, or really within, its visions of population, the field of public health necessarily retained a sense of the significance of individuals. Cumpston's primary sentiment, common to those in the position of governing with respect to health in the early to mid twentieth century, was that people *who could* should take responsibility for their own bodily condition. He concluded his paper 'Cleanliness' with this idea: 'People have developed the habit of looking to the government ... to Houses of Parliament or to Central Boards of Health to create for themselves the Kingdom of Heaven while the Kingdom is after all within themselves'.[7] Put another way, self-governance in relation to health was one of the objectives of the apparatus of public health. 'Real advance in health reforms comes only with the formation of habits. People are not really happy in doing things which require a mental effort of repeated or monotonous kind – the comfortable actions are the automatic ... we can succeed in our efforts towards health reform only so far as we can educate the people through the stage of conviction to the phase of established habits of health.'[8] At the same time, and again like many others, Cumpston subscribed fully to the idea that the state and its experts had a duty to define and identify those who could not be expected to exercise such responsibility. For them, lines of hygiene were not these interior frontiers of healthy habit and conduct, but legal and geographical segregation.

In the nineteenth century the state was under scrutiny in terms of what it could legitimately *do* to bodies of its subjects (vaccination, isolation, compulsory treatment). By the early twentieth century, there was simultaneously a sense of 'health' as an increasing right of the citizen-subject but also the emergence of a palpable sense of civic duty to *be* healthy, not only for oneself, but for oneself insofar as one's health is necessary for the nation/state or the race. Nikolas Rose's insights can in many ways be seen as a re-formulation of those offered many years ago by Cumpston:

> Liberal governmentalities will dream that the national objective for the good subject of rule will fuse with the voluntarily assumed obligations of free individuals to make the most of their own existence by conducting their life responsibly. At the same time, subjects themselves will have to make their decisions about their self-conduct surrounded by a web of vocabularies, injunctions, promises, dire warnings and threats of intervention, organized increasingly around a proliferation of norms and normativities.[9]

In my opinion this dream came closest to its realisation in the inter-war years when an intense nationalism, came together with welfare governance, and when these came together with a eugenic conception of race. We can see this dream in operation clearly in the field of public health and hygiene. This political ambition of public health often surprises, even dismays contemporary practitioners and scholars in population health, epidemiology, and other medical disciplines. But for the likes of Cumpston, the whole mission of public health was that it formed an important part of the larger modern projects of nation, of race and of colonisation.

Notes*

Introduction

1 J.H.L. Cumpston, 'Cleanliness', unpublished typescript, no date, in Cumpston Papers, National Library of Australia, Canberra MS613 Box 7 (i), p. 13.
2 Catherine Waldby, *AIDS and the Body Politic: biomedicine and sexual difference*, Routledge, 1996, p. 5.
3 The phrase 'boundaries of rule' is from Ann Laura Stoler, 'Rethinking Colonial Categories: European Communities and the Boundaries of Rule', *Comparative Studies in Society and History*, 31 (1989): 134–61.
4 For example, Maynard Swanson, 'The Asiatic Menace: Creating Segregation in Durban 1870–1900', *International Journal of African Historical Studies*, 16 (1983): 401–21; Harriet Deacon, 'Racism and Medical Science in South Africa's Cape Colony in the mid- to late Nineteenth Century', *Osiris*, 15, (2000): 190–206; Suzanne Saunders, 'Isolation: the development of leprosy prophylaxis in Australia', *Aboriginal History*, 14 (1990): 168–81; Heather Bell, *Frontiers of Medicine in the Anglo-Egyptian Sudan 1899–1940*, Clarendon Press, 1999; JoAnne Brown, 'Purity and Danger in Colour: Notes on Germ Theory and the Semantics of Segregation, 1895–1915', in Jean-Paul Gaudillière and Ilana Löwy (eds) *Heredity and Infection: The History of Disease Transmission*, Routledge, 2001, pp. 101–32; Renisa Mawani, ' "The Island of the Unclean": Race, Colonialism and "Chinese Leprosy" in British Columbia, 1891–1924', *Journal of Law, Social Justice and Global Development*, (2003) http.//elj. warwick.ac.uk/global/
5 For example, Roy Macleod and Milton Lewis (eds), *Disease, Medicine and Empire: Perpectives on Western Medicine and the Experience of European Expansion*, Routledge, 1988; David Arnold (ed.), *Imperial Medicine and Indigenous Societies*, Manchester University Press, 1988; Meghan Vaughan, *Curing their Ills: Colonial Power and African Illness*, Stanford University Press, 1991; David Arnold, *Colonising the Body: State Medicine and Epidemic Disease in Nineteenth Century India*, University of California Press, 1993; Mark Harrison, *Public Health in British India: Anglo-Indian Preventive Medicine, 1859–1914*, Cambridge University Press, 1994; Lenore Manderson, *Sickness and the State: Health and Illness in Colonial Malaya*, Cambridge University Press, 1996; Warwick Anderson, 'Immunities of Empire; Race, Disease and the New Tropical Medicine, 1900–1920', *Bulletin of the History of Medicine*, 70 (1996): 94–118; Michael Worboys, 'The Colonial World as Mission and Mandate: Leprosy and Empire, 1900–1940', *Osiris*, 15 (2000): 207–20; George Odour Ndege, *Health, State, and Society in Kenya*, University of Rochester Press, 2001; Philippa Levine, *Prostitution, Race and Politics: Policing Venereal Disease in the British Empire*, Routledge, 2003.
6 In particular, Robert Proctor's work on the medicalisation of anti-semitism and genocide in the guise of quarantine is directly relevant. See Robert N.

*Except for primary published sources, the place of publication is omitted.

Proctor, 'The Destruction of "Lives Not Worth Living" ', in Jennifer Terry and Jacqueline Urla (eds), *Deviant Bodies: Critical Perspectives on Difference in Science and Popular Culture*, Indiana University Press, 1995, pp. 170–96; Alexandra Minna Stern, 'Buildings, Boundaries and Blood: Medicalization and Nation-Building on the US-Mexico Border, 1910–1930', *Hispanic American Historical Review*, 79 (1999): 41–81; Nayan Shah, *Contagious Divides: Epidemics and Race in San Francisco's Chinatown*, University of California Press, 2002; Warwick Anderson, *The Cultivation of Whiteness: Science, Health and Racial Destiny in Australia*, Melbourne University Press, 2002.
7   A. Dirk Moses, 'Conceptual blockages and definitional dilemmas in the "racial century": genocides of indigenous peoples and the Holocaust', *Patterns of Prejudice*, 36 (2002): 7–36.
8   Nikolas Rose, 'Governing "advanced" liberal democracies', in Andrew Barry et al., *Foucault and Political Reason*, University of Chicago Press, 1996, pp. 47–50.
9   For example, Alan Petersen and Deborah Lupton, *The New Public Health: health and self in the age of risk*, Allen & Unwin, 1996; Alan Petersen, 'Risk, governance and the new public health', in Alan Petersen and Robin Bunton (eds), *Foucault, Health and Medicine*, Routledge, 1997, pp. 189–206; Sarah Nettleton, 'Governing the risky self: how to become healthy, wealthy and wise', in Ibid., pp. 207–22; Robin Bunton and Roger Burrows, 'Consumption and health in the "epidemiological" clinic of late modern medicine', in Robin Bunton, Sarah Nettleton and Roger Burrows (eds), *The Sociology of Health Promotion*, Routledge, 1995.
10 For example, Stoler, 'Rethinking Colonial Categories', pp. 136–7; Anne McClintock, *Imperial Leather: Race, Gender and Sexuality in the Colonial Contest*, Routledge, 1995, esp. pp. 368–79; Adele Perry, *On the Edge of Empire: Gender, Race and the Making of British Columbia, 1849–1871*, University of Toronto Press, 2001.
11 For a summary and reflection on these issues, see Ann Curthoys, 'Expulsion, Exodus and Exile in White Australian Historical Mythology', in Richard Nile and Michael Williams (eds), *Imaginary Homelands: The Dubious Cartographies of Australian Identity*, University of Queensland Press, 1999, pp. 1–18.
12 On race, medicine and British settlement, see Anderson *The Cultivation of Whiteness* chs 1–2.
13 Susan Craddock and Michael Dorn, 'Nationbuilding: gender, race and medical discourse', *Journal of Historical Geography*, 27 (2001): 313–18.
14 Waldby, *AIDS and the Body Politic*, p. 88.
15 Emily Martin, 'Toward an anthropology of immunology: The Body as Nation-State', *Medical Anthropology Quarterly*, 4 (1992): 410–26; Petersen and Lupton, *The New Public Health*, p. 55.
16 Nikolas Rose, *Inventing Our Selves: Psychology, Power, and Personhood*, Cambridge University Press, 1998, p. 163.
17 Robert Crawford, 'The boundaries of the self and the unhealthy other: reflections on health, culture and AIDS', *Social Science and Medicine*, 27 (1993): 1348.
18 David Sibley, *Geographies of Exclusion: Society and Difference in the West*, Routledge, 1995, p. 49; see also, Alison Bashford and Carolyn Strange, 'Isolation and Exclusion in the Modern World', in Carolyn Strange and

Alison Bashford (eds), *Isolation: places and practices of exclusion*, Routledge, 2003, pp. 1–19.

19 Ann Laura Stoler, 'Sexual Affronts and Racial Frontiers: European Identities and the Cultural Politics of Exclusion in Colonial Southeast Asia', in Frederick Cooper and Ann Laura Stoler (eds), *Tensions of Empire: Colonial Cultures in a Bourgeois World*, University of California Press, 1997, p. 199.

20 See the section 'Making Boundaries', in Cooper and Stoler (eds), *Tensions of Empire*, pp. 163–286.

21 For example, Ann Laura Stoler, 'Making Empire Respectable: The Politics of Race and Sexual Morality in Twentieth-Century Colonial Cultures', in Ann McClintock, Aamir Mufti and Ella Shohat (eds), *Dangerous Liaisons: Gender, Nation and Postcolonial Perspectives*, University of Minnesota Press, 1997, pp. 344–73. Christopher Forth, 'Moral Contagion and the Will: the crisis of masculinity in *fin-de siècle* France', in Alison Bashford and Claire Hooker (eds), *Contagion: historical and cultural studies*, Routledge, 2001, pp. 61–75.

22 Proctor, 'The Destruction of "Lives not Worth Living" ', pp. 176–9.

23 For the humanitarian and philanthropic line of public health, see, especially, Christopher Hamlin, 'State Medicine in Great Britain', in Dorothy Porter (ed.), *The History of Public Health and the Modern State*, Rodopi, 1994, p. 135.

24 Nikolas Rose, *Powers of Freedom: Reframing Political Thought*, Cambridge University Press, 1999.

25 Nikolas Rose, 'Medicine, History and the Present', in Colin Jones and Roy Porter (eds), *Reassessing Foucault: Power, Medicine and the Body*, Routledge, 1994, p. 65. See also Thomas Osborne, 'Security and vitality: drains, liberalism and power in the nineteenth century', in Andrew Barry *et al.* (eds), *Foucault and political reason: Liberalism, neo-liberalism and rationalities of government*, University of Chicago Press, 1996, pp. 99–122.

26 George Rosen, *A History of Public Health*, MD Publications, 1958, p. 110.

27 Michel Foucault, 'The Birth of Social Medicine', in James D. Faubion (ed.), *Essential Works of Michel Foucault*, Vol. 3 'Power', The New Press, 2000, pp. 137–42.

28 Rosen, *A History of Public Health*, pp. 134–5.

29 See Dorothy Porter, *Health, Civilization and the State: a history of public health from ancient to modern times*, Routledge, 1999, pp. 58–61.

30 Ibid., p. 65.

31 Graham Burchell, 'Governmental rationality', in Graham Burchell *et al.* (eds), *The Foucault Effect*, University of Chicago Press, 1991, pp. 4–5; Michel Foucault, *The History of Sexuality: an introduction*, Penguin, 1981; see also Michel Foucault, 'The Politics of Health in the Eighteenth Century', in Paul Rabinow (ed.), *The Foucault Reader*, Penguin, 1984, pp. 278–9.

32 Foucault, *The History of Sexuality; an Introduction*, p. 139.

33 Rosen, *A History of Public Health*, pp. 111–14; Ian Hacking, 'How Should We Do a History of Statistics', in Burchell *et al.* (eds), *The Foucault Effect*, pp. 181–96; see also Ray Jureidini and Kevin White, 'Life Insurance, the Medical Examination and Cultural Values', *Journal of Historical Sociology*, 13 (2000): 190–214.

34 Foucault, 'Governmentality', in Burchell *et al.* (eds), *The Foucault Effect*, p. 96.

35 Hacking, 'How Should We Do the History of Statistics?', p. 181; for a study of the significance of these forms of knowledge on public health, see John

Eyler, *Sir Arthur Newsholme and State Medicine, 1885–1935*, Cambridge University Press, 1997.
36 Mary Poovey, *Making a Social Body: British Cultural Formation, 1830–1864*, University of Chicago Press, 1995, p. 34.
37 Uday S. Mehta, 'Liberal Strategies of Exclusion', in Cooper and Stoler (eds), *Tensions of Empire*, pp. 59–86.
38 See, especially, U. Kalpagam, 'The colonial state and statistical knowledge', *History of the Human Sciences*, 13 (2000): 37–55.
39 Maureen K. Lux, *Medicine That Walks: Disease, Medicine, and Canadian Plains Native People, 1880–1940*, University of Toronto Press, 2001; see also Nicholas Thomas on 'sanitizing-colonizing', in *Colonialism's Culture: Anthropology, Travel and Government*, Polity, 1994, p. 116 ff.
40 Foucault, *Discipline and Punish*, Penguin, 1991, p. 198.
41 Ibid., p. 197.
42 Ibid., p. 198.
43 David Armstrong, 'Public Health Spaces and the Fabrication of Identity', *Sociology*, 27 (1993): 393–410.
44 Alan Sears, ' "To Teach them how to live": The Politics of Public Health from Tuberculosis to AIDS', *Journal of Historical Sociology*, 5 (1992): 70–71.
45 Deborah Lupton, *The Imperative of Health: public health and the regulated body*, Sage, 1995, pp. 10–11.
46 Ibid., p. 22.

Chapter 1

1 Margaret Pelling, 'The Meaning of Contagion: Reproduction, Medicine and Metaphor', in Alison Bashford and Claire Hooker (eds), *Contagion: historical and cultural studies*, Routledge, 2001, pp. 15–38.
2 O.A. Bushnell, *The Gifts of Civilisation: Germs and Genocide in Hawaii*, University of Hawaii Press, 1993; Sheldon Watts, Clark Spencer Larsen and George R. Milner (eds), *In the Wake of Contact: Biological Responses to Conquest*, Wiley-Liss, 1994; Margaret Jolly, 'Desire, Difference and Disease: sexual and venereal exchanges on Cook's voyages in the Pacific', in Ross Gibson (ed.), *Exchanges: cross-cultural encounters in Australia and the Pacific*, Historic Houses Trust of NSW, 1996, pp. 185–217; Maureen K. Lux, *Medicine that Walks: Disease, Medicine and Canadian Plains Native People 1880–1940*, University of Toronto Press, 2001.
3 Philip Curtin, *Death by Migration: Europe's Encounter with the Tropical World in the Nineteenth Century*, Cambridge University Press, 1989; Trevor Burnard, ' "The Countrie Continues Sicklie": White Mortality in Jamaica, 1655–1780', *Social History of Medicine*, 12 (1999): 45–72.
4 For 'isolate studies', see D.F. Roberts, N. Fujiki and K. Torizuka (eds), *Isolation, Migration and Health*, Cambridge University Press, 1992.
5 Michael Worboys discusses extensively the seed and soil metaphor in *Spreading Germs: Disease Theories and Medical Practice in Britain; 1865–1900*, Cambridge University Press, 2000.
6 See, for example, Sheldon Watts, 'Smallpox in the New World and the Old: From Holocaust to Eradication, 1518–1977', in his *Epidemics and History: Disease, Power and Imperialism*, Yale University Press, 1997, pp. 84–121;

Elizabeth A. Fenn, *Pox Americana: The Great Smallpox Epidemic of 1775–82*, Hill & Wang, 2001.
7. Margaret Pelling, *Cholera, Fever and English Medicine*, Oxford University Press, 1978, pp. 250–95; Worboys, *Spreading Germs*, pp. 40, 188–89.
8. Anne Hardy argues for the significance of isolation alongside vaccination in controlling smallpox in London. See Anne Hardy, *The Epidemic Streets: Infectious Disease and the rise of preventive medicine 1856–1900*, Oxford University Press, 1993, pp. 110–50.
9. Laura Otis, *Membranes: Metaphors of Invasion in Nineteenth-Century Literature, Science and Politics*, Johns Hopkins University Press, 1998.
10. J.Z. Bowers, 'The Odyssey of Smallpox Vaccination', *Bulletin of the History of Medicine*, 55 (1981): 17–33.
11. For Indian inoculation, see David Arnold, 'Smallpox and colonial medicine in nineteenth-century India', in David Arnold (ed.), *Imperial Medicine and Indigenous Societies*, Manchester University Press, 1988, pp. 45–65; for Montague, see Wendy Frith, 'Sex, Smallpox and Seraglios: A Monument to Lady Mary Wortley Montague', in G. Perry and M. Rossington (eds), *Femininity and Masculinity in Eighteenth Century Art and Culture*, Manchester University Press, 1994, pp. 99–122; Genevieve Miller, 'Putting Lady Mary in her Place: A Discussion of Historical Causation', *Bulletin of the History of Medicine*, 55 (1981): 2–16; see also Deborah Brunton, 'Smallpox Inoculation and Demographic Trends in Eighteenth-Century Scotland', *Medical History*, 36 (1992): 403–29.
12. The biological distinctions or relatedness of the microbes of variola and vaccinia have been disputed from Jenner's time to the present. This historic argument is both summarised and developed by the respective positions of Peter Razzell and Derrick Baxby. See Peter Razzell, *Edward Jenner's Cowpox Vaccine: the History of a Medical Myth*, Firle, 1977; Derrick Baxby, *Jenner's Smallpox Vaccine: the Riddle of the Vaccinia Virus and its Origins*, London, 1981; 'The Origins of Vaccinia Virus', comments and rejoinders in *Social History of Medicine*, 12 (1999): 139–41.
13. See, for example, J.B. Buist, *Vaccinia and Variola: a study of their life history*, J. & A. Churchill, London, 1887, pp. 1–4.
14. For compulsion, see R.M. Macleod, 'Law, medicine and public opinion: the resistance to compulsory health legislation 1870–1907', Parts I and II, *Public Law*, (1967): 107–28, 189–211; Dorothy Porter and Roy Porter, 'The politics of prevention: anti-vaccinationism and public health in nineteenth-century England', *Medical History*, 32 (1988): 231–52; Naomi Williams, 'The implementation of compulsory health legislation: infant smallpox vaccination in England and Wales, 1840–1890', *Journal of Historical Geography*, 20 (1994): 396–412; Nadja Durbach, ' "They Might as Well Brand Us": Working-Class Resistance to Compulsory Vaccination in Victorian England', *Social History of Medicine*, 13 (2000): 45–61.
15. William J. Collins, *Have you Been Vaccinated, and What Protection is it Against the Small Pox?*, H.K. Lewis, London, 1868, p. 24.
16. John Morton, *Vaccination and its Evil Consequences: Cow-pox and its Origins*, C.F. Fuller, Parramatta, 1875, p. 5.
17. Donna Haraway, 'Biopolitics of Postmodern Bodies: Constitutions of self in immune system discourse', in her *Simians, Cyborgs and Women*, Routledge, 1991, p. 204.

18 Emily Martin, *Flexible Bodies: The Role of Immunity in American Culture from the Days of Polio to the Age of AIDS*, Beacon Press, 1994.
19 See Worboys, *Spreading Germs*, pp. 120–21.
20 Evidence of Alfred Roberts, Select Committee: Opinions on Compulsory Vaccination, New South Wales Legislative Assembly, *Votes & Proceedings*, 1881, vol. 4, p. 248 (hereafter Select Committee on Vaccination, 1881).
21 Morton, *Vaccination and its Evil Consequences*, p. 5.
22 Alfred Tauber, *The Immune Self: Theory or Metaphor*, Cambridge University Press, 1994, pp. 26–7.
23 Evidence of Carl F. Fischer, Select Committee on Vaccination, 1881, p. 8.
24 J. Compton Burnett, *Vaccinosis and its cure by Thuja: with remarks on Homoeoprophylaxis*, The Homoeopathic Publishing Co., London, 1897, pp. 128–9. Of course there was no homeopathic consensus on vaccination. John le Gay Brereton was a noted Sydney homeopathic practitioner, but he entirely opposed vaccination. In 1881 he said, 'I would rather be shot than have anyone of my family vaccinated'. Evidence of John le Gay Brereton, Select Committee on Vaccination, 1881, p. 25.
25 W.D. Stokes, *Truth v. Error: A Scientific Treatise Showing the Dangers of Drugs as Medicine*, Brighton, n.d., p. 52.
26 Elizabeth Blackwell, *Scientific Method in Biology*, Ellit Stock, London, 1898, p. 65. See Mary Douglas, *Purity and Danger: An Analysis of the Concepts of Pollution and Taboo*, Routledge, 1994; Alison Bashford, *Purity and Pollution: Gender, Embodiment and Victorian Medicine*, Macmillan, 1998, p. xi.
27 E. Robinson, *Can Disease Protect Health? being a reply to Ernest Hart's pamphlet entitled 'The Truth About Vaccination'*, London, 1880.
28 J.P. Murray, *Small-pox, Chicken-Pox and Vaccination*, George Robertson, Melbourne, 1869, p. 14.
29 Evidence of John le Gay Brereton, Select Committee on the Vaccination Bill, *Journal of the NSW Legislative Council*, vol. 21 (1872): 24, p. 28 (hereafter Select Committee on Vaccination, 1872).
30 A. Beck, 'Issues in the Anti-Vaccination Movement in England,' *Medical History*, 4 (1960): 4, 313, 317. See an engraving by T. Woolnoth of 'Ann Davis' a woman with horns growing out of her head, 1806, The Wellcome Library Iconographic Collection, 46991/B. Nadja Durbach shows how working-class opposition to vaccination formed part of a political analysis of and action against a class 'tyranny'. See 'They Might as Well Brand Us', pp. 45–61.
31 Mark Harrison, *Public Health in British India*, Cambridge University Press, 1994, p. 85.
32 See Report of the Royal Commission on the Late Visitation of Small-Pox, New South Wales Legislative Assembly, *Votes & Proceedings*, vol. 2 (1883): p. 10; Porter and Porter, 'The Politics of Prevention', p. 234; Claudia Huerkamp, 'The History of Smallpox Vaccination in Germany', *Journal of Contemporary History*, 20 (1985): 628.
33 J. Beaney, *Vaccination and its Dangers*, R.N. Henningham, Melbourne, 1870, p. 11 (original emphasis).
34 A. Peripeteticus, *Cancer: A Result of Vaccination*, J.C. Stephens, Melbourne, 1898. Evidence of John le Gay Brereton, Select Committee on Vaccination, 1881, p. 28. Such theories anticipated current concerns that the appearance and virulence of Hepatitis B virus and the Human Immuno-deficiency Virus

196 Notes

in parts of Africa in the 1980s was a result of the WHO smallpox eradication campaign in the preceding decades.

35 Evidence of John T. Marx, Select Committee on Vaccination, 1872, p. 28; see also Patrick Manson, *Tropical Diseases: a Manual of Diseases of Warm Climates* [1898] Cassell, London, 1903, p. 515.
36 Jean-Paul Gaudillière and Ilana Löwy, 'Introduction: Horizontal and Vertical Transmission of Diseases: the Impossible Separation?', in Gaudillière and Löwy (eds), *Heredity and Infection: The History of Disease Transmission*, Routledge, 2001, p. 4.
37 Morton, *Vaccination and its Evil Consequences*, p. 6.
38 Christopher Hamlin, 'State Medicine in Great Britain', in Dorothy Porter (ed.), *The History of Public Health and the Modern State*, Rodopi, 1994, p. 135.
39 On public health, morality and domestic and social spaces, see Bashford, *Purity and Pollution*, ch. 1.
40 Anne Hardy, 'Smallpox In London: Factors in the Decline of the Disease in the Nineteenth Century, *Medical History*, 27 (1983): 111–38.
41 See Chapter 2; Susan Craddock, 'Sewers and Scapegoats: Spatial Metaphors of Smallpox in Nineteenth Century San Francisco', *Social Science and Medicine*, 41 (1995): 957–68; Nayan Shah, *Contagious Divides: Epidemics and Race in San Francisco's Chinatown*, University of California Press, 2001; Alan Mayne, *Fever, Squalor and Vice: Sanitation and Social Policy in Victorian Sydney*, University of Queensland Press, 1982, ch. 13; P.H. Curson, *Times of Crisis: Epidemics in Sydney 1788–1900*, Sydney University Press, 1985.
42 Anthony Wohl, *Endangered Lives: Public Health in Victorian Britain*, J.M. Dent, 1983, pp. 133–4; S.M.F. Fraser, 'Leicester and smallpox: the Leicester method', *Medical History*, 24 (1980): 315–32.
43 This is discussed in Chapter 6.
44 Cabinet of the New South Wales Government, *Council Opinions upon Compulsory Vaccination*, Government Printer Sydney, 1881, p. 223.
45 K. Walker, *The Story of Medicine*, London, Arrow, 1954, p. 230.
46 Frank Fenner, 'Smallpox: Emergence, Global Spread, and Eradication', *History and Philosophy of the Life Sciences*, 15 (1993): 397.
47 The disputed African source of HIV is a case in point here. See, for example, D. Siefkes, 'The Origin of HIV-1, The AIDS Virus', *Medical Hypotheses*, 41 (1993): 289–99. For a critique of these trends, see Douglas Haynes, 'Still the Heart of Darkness: The Ebola Virus and the Meta-Narrative of Disease in the "Hot Zone" ', *Journal of Medical Humanities*, 23 (2002): 133–45.
48 Fenn, *Pox Americana*; Lux, *Medicine that Walks*; Judy Campbell, *Invisible Invaders: smallpox and other diseases in Aboriginal Australia 1780–1880*, Melbourne University Press, 2002.
49 Lux, *Medicine that Walks*, p. 15.
50 Thomas Christie, *An Account of the ravages committed in Ceylon by Small-Pox, previously to the Introduction of Vaccination*, J&S Griffith, London, 1811, pp. 20–21. By 'inoculation' Christie here means vaccination with cowpox matter.
51 See Bowers, 'The Odyssey of Smallpox Vaccination', 17–33. For a description of Jenner's methods of distributing vaccine, see William J. Collins, *Have you Been Vaccinated, and What Protection is it Against the Small Pox?*, H.K. Lewis, London, 1868, p. vi.

52 For example, evidence of Samuel Pickford Bedford, Select Committee on Vaccination, 1872, p. 790.
53 Collins, *Have You been Vaccinated*, p. 14.
54 Evidence of Miles Egan, Select Committee on Vaccination, 1872, p. 796.
55 *Daily Telegraph* (Sydney), 14 July 1881, p. 4; *Daily Telegraph* (Sydney), 10 August 1881, p. 4.
56 See, for example, questions by Mr Deas Thomson, Select Committee on Vaccination 1872, p. 790.
57 Evidence of John le Gay Brereton, Select Committee on Vaccination, 1881, p. 27.
58 C. Creighton, *The Natural History of Cow-Pox and Vaccinal Syphilis*, Cassell, London, 1887, pp. 23–8. See also E.M. Crookshank, *The History and Pathology of Vaccination*, 2 vols, H.K. Lewis, London, 1889; H. Valentine Knaggs, *The Truth About Vaccination: The nature and origin of vaccine lymph and the teachings of the New Bacteriology*, C.W. Daniel, London, 1914.
59 This change is well documented in Mark Harrison, *Climates and Constitutions: Health, Race, Environment and British Imperialism in India 1600–1850*, Oxford University Press, 1999, esp. pp. 11–18. See also Ivan Hannaford, *Race: the history of an idea in the west*, Woodrow Wilson Centre Press, 1996.
60 Harrison, *Climates and Constitutions*, p. v.
61 Ann Laura Stoler, 'Making Empire Respectable: The Politics of Race and Sexual Morality in Twentieth-Century Colonial Cultures', in Anne McClintock, Aamir Mufti and Ella Shohat (eds), *Dangerous Liaisons: Gender, Nation and Postcolonial Perspectives*, University of Minnesota Press, 1997, pp. 344–73.
62 Christie, *An Account of the Ravages of Small-Pox*, p. 21.
63 Ibid.
64 Evidence of John le Gay Brereton, Select Committee on Vaccination, 1872, p. 27.
65 Morton, *Vaccination and its Evil Consequences*, p. 5.
66 Evidence of Charles Taylor, Select Committee on Vaccination, 1872, p. 17.
67 Evidence of John le Gay Brereton, Select Committee on Vaccination, 1872, p. 24.
68 Nikolas Rose, 'Medicine, History and the Present' in Colin Jones and Roy Porter (eds) *Reassessing Foucault: Power, Medicine and the Body*, Routledge, 1994, p. 55.
69 See Eyler, *Sir Arthur Newsholme*, p. 32.
70 Durbach, 'They Might as Well Brand Us', pp. 45–61.
71 See B.S. Cohn, *Colonialism and its Forms of Knowledge: the British in India*, Princeton University Press, 1996; Nicholas Thomas, *Colonialism's Culture: Anthropology, Travel and Government*, Polity, 1994, p. 116. See also U. Kalpagam, 'The colonial state and statistical knowledge', *History of the Human Sciences*, 13 (2000): 37–55.
72 Harrison, *Public Health in British India*, p. 82.
73 Mitchell Dean, *Governmentality: power and rule in modern society*, Sage, 1999, p. 30.
74 Christie, *An Account of the Ravages of Small-Pox*, p. 33.
75 E.S.P Bedford to Colonial Secretary, 23 February 1869, printed in NSW Legislative Assembly, *Votes & Proceedings*, 1868–9.
76 Durbach, 'They Might as Well Brand Us', p. 58.
77 See Curson, *Times of Crisis*.

78 J.H.L Cumpston and F. McCallum, *The History of Small-pox in Australia 1909–1923*, Government Printer, Melbourne, 1925, pp. 31–2, 77.
79 John Torpey, *The Invention of the Passport: Surveillance, Citizenship and the State*, Cambridge University Press, 2000. See also Chapter 5.
80 Report and Minutes, *The Australasian Sanitary Conference, 1884*, Government Printer, Sydney, 1884, p. 19.
81 Quarantine Regulations in *Canada Circular*, 12 June 1908, reprinted in *Pratique*, 25 (2000): 106.
82 A copy of such an Inspection Card, stamped 'Passed Medical and Civil Inspection' is available in V. Denis Vandervelde, 'Canada: Immigrant Health Inspection, 1909' *Pratique*, 26 (2001): 83.
83 Cumpston and McCallum, *History of Small-Pox in Australia*, p. 15.

Chapter 2

1 Alison Bashford and Carolyn Strange, 'Isolation and Exclusion in the Modern World', in Carolyn Strange and Alison Bashford (eds), *Isolation: places and practices of exclusion*, Routledge, 2003, pp. 1–19.
2 J.H.L Cumpston, *Health and Disease in Australia*, edited and introduced by Milton Lewis, Australian Government Publishing Service, 1989; Judy Campbell, *Invisible Invaders: Smallpox and other Diseases in Aboriginal Australia 1780–1880*, Melbourne University Press, 2002.
3 'An Act for the Prevention ... of the Disease called the Cholera', 2 and 3, William IV, c. 10, 1832.
4 This epidemic has been analysed for numerous purposes and from a range of epistemological positions: from early epidemiological histories produced out of the Australian bureaucracy of health itself to more recent historical geographies and social and economic histories. J.H.L. Cumpston, *The History of Small-Pox in Australia, 1788–1908*, Government Printer, Melbourne, 1914; Alan Mayne, *Fever, Squalor and Vice: Sanitation and Social Policy in Victorian Sydney*, University of Queensland Press, 1982, ch. 13; P.H. Curson, *Times of Crisis: Epidemics in Sydney 1788–1900*, Sydney University Press, 1985, ch. 6; Jean Duncan Foley, *In Quarantine: A History of Sydney's Quarantine Station 1828–1984*, Kangaroo Press, 1995, chs 5, 6; Greg Watters, 'The *S.S. Ocean*: Dealing with Boat People in the 1880s', *Australian Historical Studies*, 120 (2002): 331–43.
5 Jane Buckingham, *Leprosy in Colonial South India: medicine and confinement*, Palgrave Macmillan, 2002.
6 David Arnold, 'Touching the Body: Perspectives on the Indian Plague, 1896–1900', *Subaltern Studies*, 5 (1987): 61–3.
7 House disinfection and slum clearances, for example. See Curson, *Times of Crisis*. pp. 94–99; Mayne, *Fever, Squaler and Vice*, pp. 191–200.
8 But see Howard Markel, ' "Knocking out the Cholera": Cholera, Class and Quarantines in New York City, 1892', *Bulletin of the History of Medicine*, 69 (1995): 420–57.
9 For international examples, see Charles Rosenberg, *The Cholera Years: The United States in 1832, 1849 and 1866*, University of Chicago Press, 1962, pp. 2–3; Judith W. Leavitt, 'Politics and Public Health: Smallpox in Milwaukee, 1894–1895', *Bulletin of the History of Medicine*, 50 (1976): 553; Arnold,

'Touching the Body: Perspectives on the Indian Plague, 1896–1900', 55–90. Peter J. Tyler, 'Boards of Health: A Nineteenth Century Response to Epidemic', in Linda Bryder and Derek A. Dow (eds), *New Countries and Old Medicine: Proceedings of an International Conference of the History of Medicine and Health*, Pyramid Press, 1995.

10 Charles Rosenberg, *Explaining Epidemics and other studies in the history of medicine*, Cambridge University Press, 1992, p. 282.

11 There were a few near misses in which people on board ships off the coast, or at Thursday Island, were infected and died, but the ship was not allowed to enter the territory of the colony. The details of these events convinced politicians, public health policy-makers and doctors of the need to retain rigid maritime quarantine. See, for example, K.I. O'Doherty, 'Federal Quarantine', *Australasian Association for the Advancement of Science*, 6 (1895): 837–8.

12 Judy Campbell argues that smallpox on the Australian continent preceded British invasion, originating with Macassan fishermen on the north coast. Campbell, *Invisible Invaders*.

13 Greg Watters demonstrates how it is unlikely that the epidemic began with the Chinese community. See 'The *S.S. Ocean*', pp. 332–4.

14 For more detail, see Curson, *Times of Crisis*, pp. 104–7; Alan Mayne, 'The dreadful scourge': responses to smallpox in Sydney and Melbourne, 1881–2', in Roy Macleod and Milton Lewis (eds), *Disease, Medicine and Empire*, Routledge, 1988.

15 Foley, *In Quarantine*, pp. 36–46.

16 The linked Indigenous and colonial cultural history of this area is detailed in Maria Nugent, 'Revisiting LaPerouse: a postcolonial history', PhD thesis, University of Technology, Sydney, 2001.

17 *Sydney Morning Herald*, 16 June 1881, p. 6.

18 See 'Meeting of Cabinet – Opinions on Compulsory Vaccination', NSW Legislative Assembly, *Votes and Proceedings*, vol. 4 (1881): 1019–73; Report of the Royal Commission into Management of the Quarantine Station, NSW Legislative Assembly, *Votes and Proceedings*, 1882 (hereafter Royal Commission on Quarantine, 1882); Report of the Royal Commission on the Late Visitation of Smallpox, NSW Legislative Assembly, *Votes and Proceedings*, vol. 2, 1883 (hereafter Royal Commission on Smallpox, 1883).

19 S. Mannington Caffyn, 'The Non-Transmission of Small-Pox by Vaccine Lymph', *The Lancet*, 29 July 1893, p. 272. Mayne emphasises the commercial impact of the epidemic in 'The Dreadful Scourge'.

20 Infectious Disease Supervision Act, 1881 (NSW).

21 Stephen Garton, 'Policing the Dangerous Lunatic: Lunacy Incarceration in NSW, 1870–1914', in Mark Finnane (ed.), *Policing in Australia: Historical Perspectives*, University of New South Wales Press, 1987, pp. 74–87. For the history of policing in Australia, see Mark Finnane, *Police and Government: Histories of Policing in Australia*, Oxford University Press, 1994.

22 George Rosen, 'Cameralism and the Concept of Medical Police', *Bulletin of the History of Medicine*, 27 (1953): 21–42; George Rosen, *A History of Public Health*, MD Publications, 1958, p. 118; Paul Weindling, 'Public Health in Germany', in Dorothy Porter (ed.), *The History of Public Health and the Modern State*, Rodopi, 1994, p. 122.

23 Board of Health minutes, 25 October 1881, NSW State Archive [NSWSA] 5/2913.

24 See NSW Board of Health minutes, 18 July 1881, NSWSA, 5/2913; 'Regulations for the Establishment and Management of an Ambulance and Disinfecting Staff – Smallpox Regulations', NSW Legislative Assembly, *Votes & Proceedings*, 1881, p. 1.
25 Michel Foucault, 'The eye of power', cited in Deborah Lupton, *The Imperative of Health: Public Health and the Regulated Body*, Sage, 1995, p. 23.
26 Martin Hewitt, 'Bio-politics and Social Policy: Foucault's Account of Welfare', in Mike Featherstone *et al.* (eds), *The Body: Social Process and Cultural Theory*, Sage, 1991, pp. 234–5.
27 Mr Buchanan, *NSW Parliamentary Debates*, 1881, vol. 2, p. 2475.
28 Caffyn 'The Non-Transmission of Small-Pox', p. 272.
29 Michel Foucault, *Discipline and Punish*, Penguin, 1991, p. 197.
30 Lupton, *The Imperative of Health*, p. 64.
31 See, for example, *Sydney Morning Herald*, 10 October 1881, p. 6; 31 October 1881, p. 5; 5 November 1881, p. 5.
32 Michael Ostwald and John Moore, 'The Science of Urban Pathology: Victorian Rituals of Architectural and Urban Dissection', *Australasian Victorian Studies Journal*, 2 (1996): 65–80; Alison Bashford, *Purity and Pollution: Gender, Embodiment and Victorian Medicine*, Macmillan, 1998, ch. 1. For domestic architecture and medicine see Annemarie Adams, *Architecture in the Family Way: Doctors, Houses, and Women, 1870–1900*, McGill-Queen's University Press, 1996.
33 Michel Foucault, *The Birth of the Clinic*, Vintage, 1975, p. 31.
34 See P. Susan Hardy, ' "Surgical Spirit": Listerism in NSW', PhD thesis, University of NSW, 1990.
35 Report of the Royal Commission on Small-Pox, 1883, p. 2.
36 NSW Board of Health, Report on the Late Epidemic of Smallpox, NSW Legislative Assembly, *Votes & Proceedings*, vol. 2 (1883): 5.
37 'Medical Officer's Return – Smallpox Regulations', NSW Legislative Assembly, *Votes & Proceedings*, 1881, p. 5.
38 Report of the Royal Commission on Small-Pox, 1883, p. 1.
39 See Mayne, *Fever, Squalor and Vice*, pp. 191–207.
40 *Sydney Morning Herald*, 16 June 1881, p. 6.
41 'Rules for the guidance of surgeons visiting cases of Small-pox' in Report of the Royal Commission on Small-pox, Appendix E, NSW Legislative Assembly, *Votes and Proceedings*, vol. 2 (1883): 14.
42 Alison Bashford, 'Disinfection: from the leper colony to the operating theatre', unpublished paper.
43 A summary of this report was received by the NSW Government in 1881, and formed part of the Board of Health's Report on the Smallpox Epidemic, 1883, p. 11.
44 Mary Douglas, *Purity and Danger; an analysis of the concepts of pollution and taboo*, Routledge, 1994.
45 David Armstrong, 'Public Health Spaces and the Fabrication of Identity', *Sociology*, 27 (1993): 393–4.
46 Report of the Health Officer on the Quarantine Station, North Head, NSW Legislative Assembly, *Votes & Proceedings*, vol. 2 (1883): 2. It was important for one colony to have confidence in the efficacy of another colony's quarantine system, and detailed descriptions of the isolated nature of the

sites were often offered. Fiji's Chief Medical Officer, for example, reassured his Australasian colleagues with this description of the colony's quarantine site: 'It is surrounded by water, which is several fathoms deep everywhere save at one point, and opposite that point a guard-house is built, and is occupied by an armed guard when the station is in use ...Coolies arriving from India are, if quarantine is deemed necessary, detained on the island marked "Indian Depot" ...which is completely isolated by deep water all round. Armed guard boats are anchored at a distance of three or four hundred yards from the island, to prevent all communication'. W.McGregor to the Governor of the Crown Colony of Fiji, 27 August 1884, in Australasian Sanitary Conference, *Report and Minutes of Proceedings*, Government Printer, Sydney, 1884, p. 62

47 Report of the Board of Health upon the Late Epidemic of Smallpox, 1883, p. 3.
48 Royal Commission on Quarantine, 1882, maps appended.
49 Royal Commission on Smallpox, 1883, maps appended.
50 Report of the Health Officer on the Quarantine Station, 1883, p. 3.
51 Report of the Board of Health, 1883, reference to Map.
52 Evidence of S.M. Caffyn, Report of the Royal Commission on Quarantine, 1882, p. 1191.
53 Evidence of Constable Cook, Report of the Royal Commission on Quarantine, 1882, p. 24.
54 'Small-pox in Sydney', *Sydney Morning Herald*, 12 July 1881, p. 6.
55 Report of the Board of Health, 1883, p. 5.
56 The same isolation of the vaccinated from the unvaccinated occurred in a 1913 epidemic. 'Harshness at Quarantine' *Sydney Morning Herald*, 23 July 1914, Chief Secretary's Department, Smallpox Files, 1913–15 NSWSA, 5/5290.
57 The Principal Gaoler, Darlinghurst Gaol to the Comptroller of Prisons, 22 July 1881, in 'Vaccination in Darlinghurst Gaol', NSW Legislative Assembly, *Votes & Proceedings*, vol. 4 (1884): 5.
58 Caffyn, 'The Non-Transmission of Small-Pox', p. 272.
59 Compulsory Vaccination Act 1853 (Tasmania); Act to extend and make compulsory the practice of Vaccination, 1853 (South Australia); Act to Make Compulsory the Practice of Vaccination, 1854 (Victoria); An Ordinance to Make Compulsory the Practice of Vaccination, 1860 (Western Australia).
60 Health Act 1911 (Western Australia); Health Act 1919 (Victoria). An Act to suspend compulsory vaccination was passed in 1917 in South Australia.
61 Australasian Sanitary Conference, *Report*, p. 33.
62 Government Medical Adviser to the Colonial Secretary, 10 March 1859, NSW Legislative Assembly, *Papers*, 1858–59, p. 1033.
63 For these problems in twentieth century mass immunisation, see Claire Hooker and Alison Bashford, 'Diphtheria and Australian Public Health: bacteriology and its complex applications, c.1890–1930', *Medical History*, 46 (2002): 41–64.
64 Nadja Durbach, ' "They Might as Well Brand Us": Working-Class Resistance to Compulsory Vaccination in Victorian England', *Social History of Medicine*, 13 (2000): 50.

65 NSW Registrar-General, Report on Vaccination, NSW Legislative Assembly, *Papers*, 1856, pp. 119–21.
66 Durbach, 'They Might as Well Brand Us', pp. 58–9.
67 Caffyn, 'The Non-Transmission of Small-pox', pp. 272–3.
68 Report of the Board of Health, 1883, pp. 3–4.
69 Evidence of Geoffrey Eagar, Under-Secretary for Finance and Trade, Royal Commission on Quarantine, 1882, p. 1174.
70 For example, at a Board of Health meeting on 1 August 1881, it was resolved that the Executive Council 'should be requested to give a general authority for the Health Officer and any 2 members of the Board to take immediate action to compel the isolation or to remove or otherwise dispose of any persons whom it may be found necessary so to deal with as being likely to imperil public health'. Board of Health Minutes, NSWSA, 5/2913.
71 Board of Health Minutes, 13 January 1882, NSWSA 5/2913.
72 See for example, evidence of Inspector-General of Police, Royal Commission on Quarantine, p. 9; Superintendent of Police, Royal Commission on Quarantine, p. 18.
73 Royal Commission on Quarantine, p. 9.
74 Evidence of Stephen Mannington Caffyn, Royal Commission on Quarantine.
75 Evidence of Superintendent Read, Royal Commission on Quarantine, p. 19.
76 Evidence of Constable Cook, Royal Commission on Quarantine, p. 23.
77 Evidence of Constable Cook, Royal Commission on Quarantine, p. 24.
78 Evidence of John Hughes, Royal Commission on Quarantine, pp. 27–31.
79 'Smallpox: claims arising out of late visitation' NSW Legislative Assembly, *Votes & Proceedings*, vol 2 (1883): 950–51.
80 For plague, see Peter Curson and Kevin McCracken, *Plague in Sydney: The Anatomy of an Epidemio*, University of New South Wales Press, 1989.
81 J.H.L. Cumpston, 'Report upon the Activities of the Commonwealth Department of health from 1909–1930', typescript held in the Department of Health Library, Canberra, n.d (c.1930), n.p.
82 Ibid.

Chapter 3

1 S. Lyle Cummins, *Empire and Colonial Tuberculosis*, National Association for the Prevention of Tuberculosis, 1946, p. 8; J.W. Springthorpe, 'The Great White Plague', in *Social Sins*, The Church of England Social Questions Committee, Melbourne, 1912, p. 46.
2 H. Hyslop Thomson, *Tuberculosis and Public Health*, Longman, Green, London, 1920, p. 1.
3 Neil Thomson, 'A Review of Aboriginal Health Status', in Janice Reid and Peggy Trompf (eds), *The Health of Aboriginal Australia*, Harcourt Brace, 1997, pp. 60–61.
4 Robert Castel, 'From dangerousness to risk', in Graham Burchell *et al.* (eds), *The Foucault Effect: Studies in Governmentality*, University of Chicago Press, 1991, pp. 281–98.

5 bid., pp. 283, 286.
6 Michel Foucault, 'About the Concept of the Dangerous Individual in 19th Century Legal Psychiatry', in David. N. Weisstub (ed.), *Law and Psychiatry*, Pergamon Press, 1978, p. 17.
7 Castel, 'From dangerousness to risk', p. 283.
8 Ibid.
9 But see Claire Hooker, 'Sanitary failure and risk: pasteurisation, immunisation and the logics of prevention', in Alison Bashford and Claire Hooker (eds), *Contagion: historical and cultural studies*, Routledge, 2001, pp. 129–49.
10 John Eyler, 'Scarlet Fever and Confinement: the Edwardian Debate over Isolation Hospitals', *Bulletin of the History of Medicine*, 61 (1987): 1–24; Claire Hooker and Alison Bashford, 'Diphtheria and Australian Public Health: Bacteriology and its Complex Applications, c. 1890–1930', *Medical History*, 46 (2002): 41–64; Evelynn Hammonds, *Childhood's Deadly Scourge: the campaign to control diphtheria in New York City, 1880–1930*, Johns Hopkins University Press, 1999.
11 'The Germ Carrier and the Law', *Australasian Medical Gazette*, 19 October 1912: 407.
12 Judith Walzer Leavitt, *Typhoid Mary: Captive of the Public's Health*, Beacon Press, 1996.
13 Philippa Levine, *Prostitution, Race and Politics: Policing Venereal Disease in the British Empire*, Routledge, 2003; Philip W. Setel, Milton Lewis and Maryinez Lyons (eds), *Histories of Sexually Transmitted Diseases and HIV/AIDS in Sub-Saharan Africa*, Greenwood Press, 1999; Milton Lewis, *Thorns on the Rose: The History of Sexually Transmitted Diseases in Australia in International Perspective*, Australian Government Publishing Service, 1998.
14 See Graham Mooney, 'Public Health versus Private Practice: The Contested Development of Compulsory Infectious Disease Notification in Late-Nineteenth-Century Britain', *Bulletin of the History of Medicine*, 73 (1999): 238–67; Dorothy Porter and Roy Porter, 'The Enforcement of Health: The British Debate', in Elizabeth Fee and Daniel M. Fox (eds), *AIDS: The Burdens of History*, University of California Press, 1988, pp. 97–120; Eyler 'Scarlet Fever', pp. 1–24.
15 See Cumpston, *Health and Disease in Australia*, p. 398.
16 Alison Bashford and Carolyn Strange, 'Isolation and Exclusion in the Modern World' in Carolyn Strange and Alison Bashford (eds), *Isolation: places and practices of exclusion*, Routledge, 2003, pp. 1–19.
17 Alan Petersen and Deborah Lupton, *The New Public Health: health and self in the age of risk*, Allen & Unwin, 1996.
18 See Nikolas Rose, *Governing the Soul: The Shaping of the Private Self*, Routledge, 1990, p. 2,
19 These are detailed in S.D. Bird, 'On Chest Complaints in Australia', *Australian Medical Journal*, 12 (1867): 44–8. F.B. Smith explains these treatments in *The Retreat of Tuberculosis 1850–1950*, Croom Helm, 1988.
20 Quoted in Michael Roe, *Life over Death: Tasmanians and Tuberculosis*, Tasmanian Historical Research Association, 1999, p. 22. Original emphasis.
21 See S.D. Bird, *On Australasian Climates and their influence in the Prevention and Arrest of Pulmonary Consumption*, Longman, London, 1863; S.D. Bird, *Climate and Consumption*, Stillwell & Knight, Melbourne, 1870. Dr Bird,

himself consumptive, 'took the cure' and voyaged to Melbourne. His ideas were disputed in W. Thomson, *On Phthisis and the supposed influence of climate*, Stillwell & Knight, Melbourne, 1870. See also Linda Bryder, ' "A Health Resort for Consumptives": Tuberculosis and Immigration to New Zealand, 1880–1914', *Medical History*, 40 (1996): 453–71; J.M. Powell, 'Medical Promotion and the Consumptive Immigrant to Australia', *Geographical Review*, 63 (1973); 449–76; Warwick Anderson, *The Cultivation of Whiteness*, Melbourne University Press, 2002, pp. 57–61.

22 Linda Bryder, 'The Papworth Village Settlement – a Unique Experiment in the Treatment and Care of the Tuberculous?', *Medical History*, 28 (1984): 373.

23 Twelve institutions in 1899 and 223 by 1916. See Michael E. Teller, *The Tuberculosis Movement: A Public Health Campaign in the Progressive Era*, Greenwood Press, 1988, p. 82.

24 Katherine Ott, *Fevered Lives: Tuberculosis in American Culture since 1870*, Harvard University Press, 1996, p. 70.

25 JoAnne Brown, 'Purity and Danger in Colour: Notes on Germ Theory and the Semantics of Segregation, 1895–1915', in Jean-Paul Gaudillière and Ilana Löwy (eds), *Heredity and Infection: The History of Disease Transmission*, Routledge, 2001, pp. 101–32.

26 A.J. Proust, 'The Invalid Pension and Sickness Benefits in Australia prior to 1948', in A.J. Proust (ed.), *History of Tuberculosis in Australia, New Zealand and Papua New Guinea*, Brolga Press, 1991, p. 25; T.H. Kewley, *Social Security in Australia: Social Security and Health Benefits from 1900 to the present*, Sydney University Press, 1965, pp. 84–5; Claudia Thame, 'Health and the State in Australia', PhD thesis, Australian National University, 1974, pp. 85–114; James A. Gillespie, *The Price of Health: Australian Governments and Medical Politics, 1910–1960*, Cambridge University Press, 1991; Alison Bashford, 'Tuberculosis and economy: public health and the early welfare state', 'Tuberculosis and Economy: Public Health and Labour in the Early Welfare State', *Health and History*, 4 (2002): 19–40.

27 Michael Worboys, *Spreading Germs: Disease Theories and Medical Practice in Britain, 1865–1900*, Cambridge University Press, 2000, ch. 6.

28 Teller, *The Tuberculosis Movement*, p. 15; see also Robin Walker, 'The Struggle Against Pulmonary Tuberculosis in Australia', *Historical Studies*, 20 (1983): 443, who ascribes a causative role to Koch's discovery.

29 Linda Bryder, *Below the Magic Mountain: A Social History of Tuberculosis in Twentieth-Century Britain*, Clarendon Press, 1988, p. 23; Ott, *Fevered Lives*, pp. 53, 68.

30 'The Greatest Enemy of the Human Race: The Duty of the State', *Daily Telegraph*, 1 October 1901, newspaper cuttings on Tuberculosis, 1901–17, Mitchell Library [ML] Folio 616.2/N.

31 'Consumption: Measures for Prevention', 24 May 1906, unknown newspaper, Newspaper cuttings on Tuberculosis, 1901–17, ML Folio 616.2/N.

32 On plague, see Peter Curson and Kevin McCracken, *Plague in Sydney: The Anatomy of an Epidemic*, University of New South Wales Press, 1989.

33 John B. Trivett, *Tuberculosis in New South Wales*, William Applegate Gullick, Sydney, 1909, pp. 9–15. In New South Wales the death-rate in 1880 was 1.4 per 1,000. Overall, it had dropped by the turn of the century, the period I

am most interested in here, to 1 death per 1,000 in 1900, and 0.8 deaths by 1908. Trivett, *Tuberculosis in New South Wales*, p. 26.
34 Springthorpe, 'The Great White Plague'.
35 See Waterfall Sanatorium, Case Histories, 1909, No. 44, NSWSA, Colonial Secretary's Special Bundle, X648.
36 It was planned that there be two separate institutions there: one for curables along open-air treatment model and one for incurables, 'so that those in curable stages will be treated away and under different conditions from those in an incurable state'. 'Our Overcrowded Asylums: Selecting a Site of Consumptive Home', *Evening News*, 2 July 1906, newspaper cuttings on Tuberculosis, 1901–17, ML Folio 616.2/N.
37 For a summary of the government asylum in Australia, see Stephen Garton, *Out of Luck: Poor Australians and Social Welfare*, Allen & Unwin, 1990, pp. 54–61.
38 States of Australia, *Report on Consumption*, Government Printer, Melbourne, 1911, p. 6.
39 W. Ramsay Smith, *On Consumption*, Mason, Firth & McCutcheon, Melbourne, 1909, p. 9.
40 'The Greatest Enemy of the Human Race', *Sydney Daily Telegraph*, 1 October 1901, in Newspaper Cuttings on Tuberculosis, 1901–17, ML Folio 616.2/N.
41 States of Australia, *Report on Consumption*, p. 8. 'legal power must be taken to regulate the home-life of consumptives ...and in the case of persons who cannot or who will not take the necessary precautions at home, the decisive power of ordering them into segregation for the safety of their housemates in particular, and of the public in general'. The other recommendations were 'facilities for the collection of information' and 'the establishment of sanatoria and hospitals for advanced cases'.
42 Mitchell Dean, *Governmentality*, Sage, 1999, p. 133.
43 Springthorpe, 'The Great White Plague', p. 46.
44 Ibid., p. 46; J.H.L. Cumpston, 'Cleanliness', p. 6.
45 Deborah Lupton, *Risk*, Routledge, 1999, p. 92.
46 Nancy Tomes, *The Gospel of Germs: Women, Men and the Microbe in American Life*, Harvard University Press, 1998.
47 States of Australia, *Report on Consumption*, p. 7.
48 Ibid., p. 7. 'The qualification for admission to [a segregation hospital] must therefore be – not alone the fact of suffering from phthisis, *but the ascertained fact of living while so suffering under conditions which necessarily involve danger of infection to others.*' J. Ashburton Thompson, *On the Guidance of Public Effort Towards the Further Prevention of Consumption*, Stillwell & Co., Melbourne, 1899, p. 18 (original emphasis).
49 Sir Phillip Sydney Jones, 'Discussion Upon the Dissemination of Tuberculosis', *Australasian Association for the Advancement of Science*, 13 (1911): 701.
50 Duncan Turner, *Is Consumption Contagious?*, Melville, Mullins & Slade, Melbourne, 1894, pp. 8–9. He continued, 'When we see it fairly advocated that children should be separated from their parents, husbands from wives, brothers from sisters, and that the unfortunate victims should not kiss or even shake hands with their nearest relatives, if consumptive, surely it is time to inquire into the root of the matter', p. 13.

51 Ibid.
52 J.H.L. Cumpston, 'Report Upon the Activities of the Commonwealth Department of Health from 1909 to 1930', typescript in the Department of Health Library, Canberra, c.1930.
53 Ott, *Fevered Lives*, pp. 72–4.
54 Bryder, *Below the Magic Mountain*, p. 29.
55 'The first necessity, then, is that the consumptive be constantly, both by day and by night, in the purest possible atmosphere. Where it is feasible, send him to the mountains, to the desert, or on a long sea voyage; but tell him that, in order to obtain the greatest amount of benefit under such favourable conditions, the air which he breathes during the hours of the night should be nearly, if not quite, as pure as the atmosphere by which he is surrounded during the day.' James P. Ryan, 'The Open-Air Treatment of Phthisis', *Intercolonial Medical Congress*, 2nd Session, 1889, p. 92.
56 'The hurricane almost lifted our chalets up bodily and rain came in on every side. My pillow was wet, and spray went all over the bed-clothes. My day garments on the chair were saturated ...But, bless you, we thrive. Damp does not matter, damp does not give you cold. We walk in the rain and need not change damp clothing unless we like. The Nordrach book says one can stay all day in wet clothes and not catch cold'. Anon., *Letters from a Sanatorium*, George Robertson, Melbourne 1907, p. 15.
57 Ibid.
58 Ibid., pp. 10–11.
59 Ibid., pp. 10–11.
60 Erving Goffman, *Asylum: Essays on the Social Situation of Mental Patients and Other Inmates*, Penguin, 1991.
61 Department of Public Health, Victoria, *Greenvale Sanatorium for Consumptives*, Government Printer, Melbourne, 1912, pp. 8–9.
62 F.J. Drake, 'Sanatorium Treatment of Pulmonary Tuberculosis', *Australasian Medical Congress Transactions*, Session 8 (1908): 153.
63 Anon., *Letters from a Sanatorium*, pp. 10–11.
64 Ibid., p. 21.
65 Department of Public Health, Victoria, *Greenvale Sanatorium for Consumptives*, pp. 11–19.
66 Ibid., p. 19.
67 See, for example, Michael Rosenthal, *The Character Factory: Baden-Powell's Boy Scouts and the Imperatives of Empire*, Pantheon, 1984.
68 Sir Phillip Sydney Jones, 'Discussion Upon the Dissemination of Tuberculosis', *Australasian Association for the Advancement of Science*, 13 (1911): 698–9.
69 'Dr Trudeau considers that the principle aim of the modern sanatorium treatment of tuberculosis (phthisis) is to improve the patient's condition and increase his resistance to the disease by placing him under the most favourable environment obtainable. The main elements of such an environment are an invigorating climate, an open-air life, rest, coupled with the careful regulation of the daily habits, and an abundant supply of nutritious food ...The line of treatment consists in rest out of doors in all weathers, the patients being well wrapped up. Constant exposure at all temperatures, and in all weathers.' A.H. Gault, 'A Plea for the Sanatorium Treatment of Consumption', *Intercolonial Medical Congress of Australasia*, 1902, p. 514.

70 Charles Bage, 'The Treatment of Consumptives in Private Practice', *Australasian Medical Congress Transactions*, 1 (1911): 221.
71 Aldridge Evelyn, 'A Cure for Tuberculosis', *The Lone Hand*, 1 August 1911, p. 312.
72 Gault, 'A Plea for the Sanatorium Treatment of Consumption', p. 515. The Greenvale Sanatorium in Victoria said that its primary purpose was 'to develop resistance of the body to the disease, to arrest the progress of the diseases, and to restore the patient to his normal condition'. Department of Public Health, *Greenvale Sanatorium for Consumptives*, p. 4.
73 Anon., *Letters from a Sanatorium*.
74 Sir James Kingston, cited in J. Gordon Hislop, 'The Control of Pulmonary Tuberculosis: Sanatorium Treatment', *Medical Journal of Australia*, 31 May 1924, p. 531.
75 Hislop, 'The Control of Pulmonary Tuberculosis: Sanatorium Treatment', p. 529.
76 Anon., *Letters from a Sanatorium*, p. 9.
77 Ibid., p. 59.
78 Ibid., p. 59.
79 Ibid., pp. 59–60.
80 Rose, *Governing the Soul*, p. 10.
81 John Dale, 'Publicity in Public Health Administration', *Health*, 1 (1923): 57.
82 States of Australia, *Report on Consumption*, p. 7.
83 Rose, *Governing the Soul*, p. 2.
84 Ramsay Smith, *On Consumption*, p. 9.
85 The Commonwealth of Australia, Department of Trade and Customs, Committee Concerning Causes of Death and Invalidity in the Commonwealth, *Report on Tuberculosis*, Government Printer, 1916, p. 32.
86 Bage, 'The Treatment of Consumptives in Private Practice', p. 221.
87 Anon., *Letter from a Sanatorium*, p. 13.
88 Ibid., pp. 61–2, 70.
89 See Chapter 7 and the NSW Department of Health Poster 'Sow the Seeds of Good Health', Figure 7.1
90 Department of Public Health, *Greenvale Sanatorium for Consumptives*, p. 5.

Chapter 4

1 For example, Susan Burns, 'From "Leper Villages" to Leprosaria: Public Health, Nationalism and the Culture of Exclusion in Japan', in Carolyn Strange and Alison Bashford (eds), *Isolation: places and practices of exclusion*, Routledge, 2003, pp. 104–18.
2 Suzanne Saunders, 'Isolation: the Development of Leprosy Prophylaxis in Australia', *Aboriginal History*, 14 (1990): 168–81; Harriet Deacon, 'Leprosy and Racism at Robben Island', *Studies in the History of Cape Town*, 7 (1994): 45–83; Sheldon Watts, *Epidemics and History: Disease, Power and Imperialism*, Yale University Press, 1997; R.D.K. Herman, 'Out of sight, out of mind, out of power: leprosy, race and colonization in Hawai'i', *Journal of Historical Geography*, 27 (2001): 319–37; Jane Buckingham, *Leprosy in Colonial South India: medicine and confinement*, Palgrave Macmillan, 2002; Renisa Mawani, ' "The Island of the Unclean": Race,

Colonialism and "Chinese Leprosy" in British Columbia, 1891–1924', *Journal of Law, Social Justice and Global Development* (2003) http.//elj.warwick.ac.uk/global/
3   Patrick Manson, *Tropical Diseases: A Manual of the Diseases of Warm Climates*, [1898] Cassell, 1903, pp. 480–81.
4   This is Zachary Gussow's argument in *Leprosy, Racism and Public Health: Social Policy in Chronic Disease Control*, Westview Press, 1989.
5   H.P. Wright, *Leprosy: An Imperial Danger*, J. & A. Churchill, London, 1889.
6   Megan Vaughan, *Curing Their Ills: Colonial Power and African Illness*, Polity, 1991, ch. 4; Harriet Deacon, 'A history of the medical institutions on Robben Island, 1846–1910', DPhil thesis, University of Cambridge, 1994, ch. 6; Sanjiv Kakar, 'Leprosy in British India, 1860–1940: Colonial Politics and Missionary Medicine', *Medical History*, 40 (1996): 215–30; Michael Worboys, 'The Colonial World as Mission and Mandate: Leprosy and Empire 1900–1940', *Osiris*, 15 (2001): 207–18.
7   Michel Foucault, *Discipline and Punish*, Penguin, 1991, p. 198.
8   Warwick Anderson, 'Leprosy and Citizenship', *Positions*, 1998 (6): 707–29.
9   For other instances of the regulation of race through health, and health through race, see Harriet Deacon, 'Racial Segregation and Medical Discourse in Nineteenth Century Cape Town, *Journal of Southern African Studies*, 22 (1996): 187–308; JoAnne Brown, 'Purity and Danger in Colour: Notes on Germ Theory and the Semantics of Segregation, 1895–1915', in Jean-Paul Gaudillière and Ilana Löwy (eds), *Heredity and Infection: The History of Disease Transmission*, Routledge, 2001, pp. 101–32.
10 The Mission to Lepers in India was founded in 1874, a National Leprosy Fund was created in 1889 after the death of Father Damien in Hawaii, and in 1923 the British Empire Leprosy Relief Association was formed. See Buckingham, *Leprosy in Colonial South India*, pp. 152–4.
11 This debate is detailed authoritatively in Michael Worboys, 'An Imperial Danger': Leprosy and Contagion, 1860–1900' unpublished paper. My thanks to Michael Worboys for sharing this paper, and for discussion on leprosy.
12 Thanks to Harriet Deacon for discussion on these points.
13 Royal College of Physicians, *Report on Leprosy*, George Eyre and William Spottiswoode, London, 1867.
14 Charles Bruce, 'Mr Chamberlain and the Health of the Empire', *The Empire Review*, 8 (1905): 108–21. On tropical medicine and late nineteenth-century imperial politics, see Douglas M. Haynes, *Imperial Medicine: Patrick Manson and the Conquest of Tropical Disease*, University of Pennsylvania Press, 2001.
15 Agnes Lambert, 'Leprosy: Present and Past', *The Nineteenth Century*, 16 (1884): 212.
16 Alison Bashford and Claire Hooker, 'Disinfecting Mail – from smallpox to anthrax', in Alison Bashford and Claire Hooker (eds), *Contagion: epidemic, history and culture from smallpox to anthrax*, Pluto Press, 2002, pp. 227–31 (revised edn).
17 Margaret Pelling, 'The Meaning of Contagion: reproduction, medicine and metaphor', in Bashford and Hooker (eds), *Contagion*, pp. 29–32; Michael Worboys, *Spreading Germs: Disease Theories and Medical Practice in Britain, 1865–1900*, Cambridge University Press, 2000, pp. 28–42.

18 For discussion on this point, see P. Baldwin, *Contagion and the State in Europe, 1830–1930*, Cambridge University Press, 1999.
19 Royal College of Physicians, *Report on Leprosy*, p. v.
20 Lambert, 'Leprosy', p. 222.
21 The responses to the question about contagion are collated in the *Report*, pp. xliii–xlv. For changing understandings of leprosy and contagion, see Deacon, 'The history of medical institutions', ch. 6; Buckingham, *Leprosy in Colonial South India*, ch. 1; Worboys, 'An Imperial Danger'.
22 Royal College of Physicians, *Report on Leprosy*, p. lxix.
23 Lambert, 'Leprosy', p. 218.
24 Ibid., p. 217.
25 For competing and simultaneous rationales of isolation, see Alison Bashford and Carolyn Strange, 'Isolation and Exclusion in the Modern World', in Strange and Bashford, *Isolation*, p. 1ff.
26 Buckingham argues that in colonial South India, leprosy confinement was largely about the management of vagrancy, poverty and criminality. See *Leprosy in Colonial South India*, pp. 36–60.
27 Worboys, 'The Colonial World as Mission and Mandate', p. 213.
28 Royal College of Physicians, *Report on Leprosy*; p. xlv.
29 Ibid., p. xlvi.
30 Ibid., p. 203.
31 See Zachary Gussow and George S. Tracy, 'Stigma and the Leprosy Phenomenon: The Social History of a Disease in the Nineteenth and Twentieth Centuries', *Bulletin of the History of Medicine*, 44 (1970): 435.
32 Robert Koch's Nobel Lecture, 19 December 1905, is available at Nobel *e*-Museum, <http://www.nobel.se/medicine/laureates/1905/koch-lecture.html>.
33 Harriet Deacon, 'Racism and Medical Science in South Africa's Cape Colony', *Osiris*, 12 (2000): 204.
34 Kakar, 'Leprosy in British India', pp. 220–21; Worboys, 'The Colonial World as Mission and Mandate', p. 213; Buckingham, *Leprosy in Colonial South India*, ch. 7.
35 Megan Vaughan compares colonial African leper colonies and the British sanatorium in *Curing Their Ills*, pp. 95–7.
36 Brochure for the British Empire Leprosy Relief Association, n.d., Sir Leonard Rogers Papers, Wellcome Library for the History and Understanding of Medicine, PP/ROG C13 Series III 157.
37 Wright, *Leprosy*, p. 8.
38 Lambert, 'Leprosy', p. 214; Mawani, ' "The Island of the Unclean"'.
39 External Affairs memo, c. 1908, 'Leprosy in The Commonwealth', National Archives of Australia [NAA] A1 1908/4507. See also Mawani, 'The Island of the Unclean'.
40 Frederick Jones to the Secretary, Minister of External Affairs, 31 July 1907, 'Leprosy in The Commonwealth', NAA A1 1908/4507,
41 According to a document from 1925, these powers were exercised in the West Indian colonies, the Malay States, Ceylon, Malta and 'most of the larger central African colonies'. 'Memorandum on the Prevalence and Prophylaxis against leprosy in the British Empire', p. 9, 1925, Sir Leonard Rogers Papers, PP/ROG C13 Series III 222.
42 Worboys, 'An Imperial Danger'.

43 'Correspondence and Papers on Leprosy Investigations' Sir Leonard Rogers Papers, PP/ROG C13/Series I 36.
44 'The War on Leprosy' by Major-General Sir Leonard Rogers, n.d. newscuttings, Sir Leonard Rogers Papers, PP/ROG C. 13 Series I/5.
45 Leonard Rogers, *The Foundation of the British Empire Leprosy Relief Association and its first 21 years of Work*, pp. 1–2, Sir Leonard Rogers' Papers, PP/ROG C 13 Series III
46 Worboys, 'The Colonial World as Mission and Mandate', pp. 215–16.
47 Vaughan, *Curing Their Ills*, pp. 82, 89–92; Anderson, 'Leprosy and Citizenship', p. 708.
48 R. Tennyson Allan, 'Leprosy at Nauru, Central Pacific', Doctor of Medicine Thesis, University of Melbourne, 1939, p. 39. Copy in Burkitt-Ford Library, University of Sydney.
49 W.A. Newman, Administrator of the Mandated Territory of Nauru to Leonard Rogers 29 September 1930; Leonard Rogers to P.E. Deane, Prime Minister's Department, 29 April 1925, Sir Leonard Rogers' Papers, PP/ROG C13 Series V, 534.
50 Allan, 'Leprosy at Nauru', pp. v, 76.
51 W.A. Newman, Administrator of the Mandated Territory of Nauru to Leonard Rogers 29 September 1930, Sir Leonard Rogers' Papers, PP/ROG C13 Series V, 534.
52 Jeremy Bentham, *Works* (1843) cited in Michel Foucault, *Discipline and Punish*, Penguin, 1977, pp. 206–07.
53 'The Leper Home and Hospital, Dichpali', Administration Report 1932 in Sir Leonard Rogers' Papers, PP/ROG C13 Series VIII/731.
54 Dr G.A. Ryrie, *The Leper Settlement at Sungei Buloh in the Federated Malay States*, Malaya Publishing House, Singapore, 1933, p. 5.
55 Alan Petersen, 'Risk, governance and the new public health', in Alan Petersen and Robin Bunton (eds), *Foucault, Health and Medicine*, Routledge, 1997, p. 194.
56 Uday S. Mehta, 'Liberal Strategies of Exclusion', in Frederick Cooper and Ann Laura Stoler (eds), *Tensions of Empire: Colonial Cultures in a Bourgeois World*, University of California Press, 1997, p. 59.
57 Sir Leonard Rogers, 'Recent Progress in the Treatment of Leprosy and its Bearing on Prophylaxis', *Proceedings of the Pan-Pacific Science Congress*, 2 (1923): 1418.
58 L. Rogers, 'When Will Australia Adopt Modern Prophylactic Measures Against Leprosy?', *Medical Journal of Australia*, 18 (1930): 525–7. See also L. Rogers, 'Recent Progress in the Treatment of Leprosy and its Bearing on Prophylaxis', pp. 1410–18. Rogers' annoyance is evident in Leonard Rogers to the editor of the *Medical Journal of Australia*, 11 May 1930, Sir Leonard Rogers' Papers, PP/ROG C13 Series VIII/731
59 For the use of islands to segregate different kinds of problem populations, see Raymond Evans, 'The Hidden Colonists: Deviance and Social Control in Colonial Queensland', in Jill Roe (ed.), *Social Policy in Australia*, Cassell, 1976, pp. 74–100.
60 Peter Ludlow, *The Exiles of Peel Island*, Stones Corner, 1991; J. Macguire, 'The Fantome Island Leprosarium', in Roy MacLeod and Donald Denoon (eds), *Health and Healing in Tropical Australia and Papua New Guinea*,

James Cook University Press, 1991; see also Douglas Lush, John C. Hargrave and Angela Merianos, 'Leprosy Control in the Northern Territory', *Australian and New Zealand Journal of Public Health*, 22 (1998): 709–13.
61   See Raymond Evans, Kay Saunders and Kathryn Cronin, *Exclusion, Exploitation and Extermination: Race Relations in Colonial Queensland*, Australia and New Zealand Book Co., 1975, pp. 302–7.
62   Cecil Cook, *The Epidemiology of Leprosy in Australia: being the report of an investigation in Australia during the years 1923–1925*, Government Printer, Canberra, 1927, p. 9. Cecil Cook was born in 1897 and studied medicine as well as anthropology at the University of Sydney. In 1923 he received a research scholarship from the London School of Hygiene and Tropical Medicine. His appointment as Chief Protector, Quarantine Officer and Chief Medical Officer was in 1927. For studies of Cook's work as Chief Protector, see Andrew Markus, *Governing Savages*, Allen & Unwin, 1990, ch. 6; Tony Austin, *Never Trust a Government Man: Northern Territory Aboriginal Policy 1911–1939*, Northern Territory University Press, 1997, chs 6–8.
63   E.H. Molesworth, *Medical Journal of Australia*, 12 March 1927, p. 388.
64   J. Ashburton Thompson, 'A Contribution to the History of Leprosy in Australia', in *Prize Essays on Leprosy*, The New Sydenham Society, London, 1897, p. 206.
65   Ibid., pp. 108–9.
66   Police Report, Cooktown, 1 October 1903, Queensland State Archives (QSA) COL 266.
67   Sergeant of Police to the Commissioner of Police, 4 September 1900, QSA COL 265.
68   Nipper Tabagee personal communication to Ernest Hunter. *Aboriginal Health and History: Power and prejudice in remote Australia*, Cambridge University Press, 1994, p. 64.
69   J. Ashburton Thompson, 'Is Leprosy a Telluric Disease', *Australasian Association for the Advancement of Science*, 6 (1895): 786.
70   Ibid., p. 786.
71   E.H. Molesworth to Mrs Brown, 9 November 1937, Queensland Home Secretary's Office (QHSO), QSA, COL 323.
72   Dr J.H. Vivian Ross to Queensland Home Secretary, 2 June 1924, QHSO, QSA, COL 324.
73   J.H.L. Cumpston, *Health and Disease in Australia* (Milton Lewis, ed.) Australian Government Publishing Service, 1989, p. 9. p. 217.
74   Cook, *Epidemiology of Leprosy*, p. 298.
75   Ibid.
76   'Leprosy in Australia', Newsclipping, 12 April 1934, QHSO, QSA, COL 324.
77   Raphael Cilento, 'Brief Review of Leprosy in Australia and its Dependencies', Appendix 3, Report of the Seventh Session, Federal Health Council, 1934.
78   A.H. Humphry, 'Leprosy among Full-Blooded Aborigines of the Northern Territory', *Medical Journal of Australia*, 26 April 1952, p. 571.
79   'Leprosy and its Management', *Health*, 8 (1958): 21.
80   See for example Central Board of Health to Home Secretary, 20 December 1904, QSA COL 266.

81 J.S.C. Elkington to Under-Secretary, Home Secretary's Department, 17 January 1910, QHSO, QSA COL 322.
82 Alison Bashford, 'Female Bodies at Work: Gender and the Re-forming of Colonial Hospitals', *Australian Cultural History*, 13 (1994): 65–81.
83 Anon. to Home Secretary, 19 May 1908, QHSO, QSA COL 322.
84 Ibid.
85 Dunwich Inmates to Home Secretary, 4 October 1899. Correction in original.
86 J.S.C. Elkington to Under-Secretary, Home Secretary's Department, 17 January 1910, QHSO, QSA COL 322.
87 Inmates of Peel Island to Governor Sir Leslie Wilson, 27 September 1939, QHSO, QSA COL 323.
88 Hon. Secretary Peel Island Welfare Association to Queensland Premier, 29 February 1940, QHSO, QSA, COL 323. Cilento responded: 'The request that compulsory segregation be abandoned cannot possibly be accepted, particularly as leprosy is still uncontrolled in this State'. Cilento to the Under-Secretary, 13 October 1939, QHSO, QSA, COL 323.
89 A. Dodson to R. Bedford MLA, 17 November 1938, QHSO, QSA, COL 323.
90 The White Lepers of Peel Island to the Home Secretary, 2 January 1920, QHSO, QSA COL 323.
91 Alan Petersen and Deborah Lupton, *The New Public Health: health and self in the age of risk*, Allen & Unwin, 1996, pp. 61–88.
92 Mitchell Dean, *Governmentality*, Sage, 1999, pp. 135, 133.
93 Saunders, 'Isolation', pp. 168–81; see also Evans, 'The Hidden Colonists', pp. 74–100.
94 J. Chesterman and B. Galligan, *Citizens Without Rights: Aborigines and Australian Citizenship*, Cambridge University Press, 1997, p. 57.
95 For Cilento, this research demonstrated the infectiousness of leprosy and thus supported his commitment to compulsory isolation. See Minister for Health and Home Affairs to Mr Bedford MLA, 8 December 1938, QHSO, QSA, COL 323.
96 See C.E. Cook, Report on Tuberculosis – Alice Springs, 19 December 1934, NAA A431 1949/422.
97 C.E. Cook cited in 'Control of Tuberculosis Among Natives in the Northern Territory', 29 December 1953, NAA 1431 1949/422.
98 F.J.S. Wise, Incidence and Control of Tuberculosis Among Natives in the Northern Territory. Report to the Secretary, Department of Territories, 9 April 1954, NAA A431 1949/422.
99 Cook, *Epidemiology of Leprosy*, p. 63. See also Vaughan, 'Curing Their Ills', p. 81.
100 Raphael Cilento, *The White Man in the Tropics*, Government Printer, Melbourne, 1925, p. 57.
101 *Tropical Australia: Report of the Discussion at the Australasian Medical Congress, 1920*, Government Printer, Melbourne, 1921, p. 5.
102 C.E. Cook, 'Leprosy in Australia' Appendix F, in J.H.L. Cumpston, 'Report upon the Activities of the Commonwealth Department of Health from 1909–1930'. Typescript in Department of Health Library, Canberra, c.1930, n.p.

103 E.H. Molesworth, *Medical Journal of Australia*, 12 March 1927, p. 389.
104 Ibid.
105 Section 2, Native Administration Amendment Act, 1941 (WA).
106 Section 2a, 2b, 2c, 2d. Native Administration Amendment Act, 1941 (WA). See also Mary Anne Jebb, *Blood, Sweat and Welfare: A History of White Bosses and Aboriginal Pastoral Workers*, University of Western Australia Press, 2002, pp. 136–66; Hunter, *Aboriginal Health and History*, p. 39, p. 67.
107 Ann Laura Stoler, *Race and the Education of Desire*, Duke University Press, 1996, p. 52.
108 Ann Laura Stoler, 'Making Empire Respetctable: The Politics of Race and Sexual Morality in Twentieth-Century Colonial Cultures', in A. McClintock, A. Mufti and E. Shoat (eds), *Dangerous Liaisons: Gender, Nation and Postcolonial Perspectives*, Minneapolis, University of Minnesota Press, 1997, p. 360.
109 See 'Leprosy', *Australian Medical Journal*, 11 (1889): 383–6; Vaughan, *Curing Their Ills*, p. 82.
110 Deacon, 'A history of the medical institutions', ch. 6.
111 Royal College of Physicians, *Report on Leprosy*, pp. xliii–xlv.
112 The Australasian Sanitary Conference, *Report and Minutes of Proceedings*, Government Printer, Sydney, 1884, pp. 17–25.
113 See Evans, Saunders and Cronin, *Exclusion, Exploitation and Extermination*, pp. 302–7.
114 Cook, *Epidemiology of Leprosy*, p. 20.
115 Cook, 'Leprosy in Australia', in Cumpston, 'Report', n.p.
116 Cook, *Epidemiology of Leprosy*, p. 4.
117 Cook, 'Leprosy in Australia', in Cumpston, 'Report', n.p.
118 Cook, 'Report' Health Department of Western Australia, File 888/1923, cited in Hunter, *Aboriginal Health and History*, p. 61.
119 These changing policies and practices are examined in detail and from different perpectives by Peter Read, *The Stolen Generations: The Removal of Aboriginal People in NSW 1883 to 1969*, NSW Ministry of Aboriginal Affairs, 1981; Tony Austin, 'Cecil Cook, Scientific Thought and "Half-Castes" in the Northern Territory 1927–1939', *Aboriginal History*, 14 (1990): 104–22; Russell McGregor, *Imagined Destinies: Aboriginal Australians and the Doomed Race Theory, 1880–1939*, Melbourne University Press, 1997; *Bringing Them Home: Report of the National Inquiry into the Separation of Aboriginal and Torres Strait Islander Children from their Families*, Human Rights and Equal Opportunity Commission, Sydney, 1997; Fiona Paisley, *Loving Protection?: Australian Feminism and Aboriginal Women's Rights, 1919–1939*, Melbourne University Press, 2000; Warwick Anderson, *The Cultivation of Whiteness*, Melbourne University Press, 2002; pp. 216–43; Anna Haebich, *Broken Circles: Fragmenting Indigenous Families 1800–2000*, Fremantle Arts Centre Press, 2000.
120 Patricia Jacobs, 'Science and Veiled Assumptions: Miscegenation in Western Australia, 1930–1937', *Australian Aboriginal Studies*, 2 (1986): 15–23.
121 Cecil Cook in *Aboriginal Welfare: Initial Conference of Commonwealth and State Aboriginal Authorities*, Government Printer, 1937, pp. 13–17.
122 Ibid., p. 34.
123 Cook, *Epidemiology of Leprosy*, p. 20.

124 See Anne McClintock, *Imperial Leather: Race, Gender and Sexuality in the Imperial Contest*, Routledge, 1995, pp. 352–89.
125 Anderson, *The Cultivation of Whiteness*, pp. 175–6.
126 Etienne Balibar, 'Fichte and the Internal Border', in *Masses, Classes, Ideas: studies on Politics and Philosophy Before and After Marx*, trans. James Swenson, Routledge, 1994, p. 63.
127 Ann Laura Stoler, 'Sexual Affronts and Racial Frontiers: European Identities and the Cultural Politics of Exclusion in Colonial Southeast Asia', in Cooper and Stoler (eds), *Tensions of Empire*, p. 199.
128 Cook, *Epidemiology of Leprosy*, p. 62.
129 See, for example, Cilento, *The White Man in the Tropics*, pp. 75–92; Phyllis Cilento and Raphael Cilento, 'The Mother and the Child in the Tropics of the Austra-Pacific Zone', no date, Cilento Papers, Fryer Library, Queensland, MSS 44/137.
130 Cilento, *The White Man in the Tropics*, pp. 75–92; A.T. Yarwood, 'Sir Raphael Cilento and the White Man in the Tropics', in Roy Macleod and Donald Denoon (eds), *Health and Healing in Tropical Australia and Papua New Guinea*, James Cook University Press, 1991, pp. 47–63;
131 Raphel Cilento, Speech to the New Settlers League, 11 October 1935, Cilento Papers, Fryer Library University of Queensland MSS 44/93.
132 Cilento, *The White Man in the Tropics*, p. 4.
133 Cook, *Epidemiology of Leprosy*, p. 93.
134 See Anderson, *The Cultivation of Whiteness*, passim.
135 Cook in *Aboriginal Welfare*, pp. 17–18.
136 Cook, 'Leprosy in Australia', in Cumpston, 'Report', n.p.
137 Although there were numerous statutes that prohibited marriage between white men and Aboriginal women. One practitioner working in Aboriginal health in the 1950s and 1960s called this 'Apartheid in Australia'. See Charles Duiguid, *Doctor and the Aborigines*, Rigby, 1972, pp. 181–93.
138 Stoler, 'Sexual Affronts and Racial Frontiers', p. 199.
139 J.H.L. Cumpston, 'Cleanliness', n.d. Unpublished typescript in Cumpston Papers, National Library, Canberra, MS613 Box 7 (i).

Chapter 5

1 Winichakul Thongchai, *Siam Mapped: A History of the Geo-Body of a Nation*, University of Hawaii Press, 1994, pp. 16–17.
2 Exceptions are Alexandra Minna Stern, 'Buildings, Boundaries and Blood: Medicalization and Nation-Building on the US-Mexico Border, 1910–1930', *Hispanic American Historical Review*, 79 (1999): 41–81; Nayan Shah, *Contagious Divides: Epidemics and Race in San Francisco's Chinatown*, University of California Press, 2001; Warwick Anderson, *The Cultivation of Whiteness: Science, Health and Racial Destiny in Australia*. Melbourne University Press, 2002.
3 J.H.L. Cumpston and F.F. McCallum, *The History of Smallpox in Australia 1909–1923*, Government Printer, Melbourne, 1925, p. 1.

4 Alan Petersen and Deborah Lupton, *The New Public Health: health and self in the age of risk*, Allen & Unwin, 1996, pp. 64–72.
5 W.F. Bynam, 'Policing Hearts of Darkness: Aspects of the International Sanitary Conferences', *History and Philosophy of the Life Sciences*, 15 (1993): 427.
6 For international conferences, see N. Howard-Jones, *The Scientific Background of the International Sanitary Conferences, 1851–1938*, World Health Organization, 1975; Paul Weindling, 'Introduction: constructing international health between the wars', in Weindling (ed.), *International Health Organisations and Movements, 1918–1939*, Cambridge University Press, 1995, pp. 1–16. See also Ronald E. Coons, 'Steamships and Quarantines at Trieste, 1837–1848', *Journal of the History of Medicine and Allied Sciences*, 44 (1989): 28–55.
7 Mark Harrison details the significance of pilgrimage, trade and quarantine for British governance in India in *Public Health in British India*, Cambridge University Press, 1994, pp. 117–38.
8 Alan Mayne, ' "The dreadful scourge": responses to smallpox in Sydney and Melbourne, 1881–2', in Roy Macleod and Milton Lewis (eds), *Disease, Medicine and Empire: Perspectives on Western Medicine and the Experience of European Expansion*, Routledge, 1988.
9 Australasian Sanitary Conference, Sydney, *Report and Minutes of Proceedings*, Government Printer, Sydney, 1884, pp. 12–13.
10 Thomas Borthwick, *Quarantine*, Vardon & Pritchard, Adelaide, 1901, p. 4.
11 Bynam, 'Policing Hearts of Darkness', p. 434.
12 Charles Mackellar, 'Federal Quarantine', *Journal of the Royal Society of New South Wales*, 17 (1883): 284.
13 Borthwick, *Quarantine*. See also 'The Passing of Quarantine', *Australasian Medical Gazette*, 20 April 1904: 167
14 John Torpey, *The Invention of the Passport: Surveillance, Citizenship and the State*, Cambridge University Press, 2000, p. 1.
15 On pp. 102–03, Torpey, ibid., discusses the eugenic and race-based health requirements of US entry in the early twentieth century,
16 K.I. O'Doherty, 'Federal Quarantine', *Proceedings of the Australasian Association for the Advancement of Science*, 6 (1895): 836.
17 Helen Irving, *To Constitute a Nation: A Cultural History of Australia's Constitution*, Cambridge University Press, 1999, pp. 1–5.
18 The Australasian Sanitary Conference, Sydney, 1884; The Australasian Quarantine Conference, Melbourne, 1896; The Australian and Tasmanian Intercolonial Plague Conference, Melbourne, 1900; The Commonwealth of Australia Quarantine Conferences 1904 and 1909.
19 Charles K. Mackellar, 'Federal Quarantine' quoted in J.H.L. Cumpston, 'Report upon the Activities of the Commonwealth Department of Health from 1909–1930'. Typescript in Department of Health Library, Canberra.
20 Australasian Sanitary Conference, *Report*, p. 53.
21 Although I focus here on national-level government, the implementation of health measures resides rather more typically with local administrative bodies. Weindling suggests this of Weimar and to some extent Nazi Germany, and Anne Hardy argues the same of England. See Paul Weindling, 'Public Health in Germany', in Dorothy Porter (ed.), *The History of Public Health and the Modern State*, Rodopi, 1994, p. 119. Anne Hardy, *The*

Epidemic Streets: Infectious Disease and the Rise of Preventive Medicine, 1856–1900, Clarendon Press, 1993, p. 4.
22 Michael Roe, 'The Establishment of the Australian Department of Health: Its Background and Significance', *Historical Studies*, 17 (1976): 176–92.
23 J.H.L. Cumpston, 'Report upon the Activities of the Commonwealth Department of Health' Appendix A: the Evolution of Quarantine (unpaginated).
24 Cumpston, 'The Evolution of the Department of Health', in 'Report upon the Activities of the Commonwealth Department of Health'; Anthea Hyslop, 'A Question of Identity: J.H.L. Cumpston and Spanish Influenza, 1918–1919', *Australian Cultural History*, 16 (1997/98): 60–76.
25 Cumpston, 'The Evolution of the Department of Health', Section 6, in 'Report upon the Activities of the Commonwealth Department of Health'. Milton Lewis has summarised the trajectory of this proposed expansion well. A new Commonwealth Department of Health might 'concern itself with investigation of causes of disease and death, methods of prevention, collection of data, and education of the public ...interest in national and international communication of disease ...a national system of antenatal clinics and maternity wards; the Commonwealth could subsidise State efforts to control disease, directly conduct preventive campaigns where a number of States was involved, and generally coordinate measures without infringing State sovereignty; a commitment to research could evolve out of the Commonwealth's existing facilities – the Australian Institute of Tropical Medicine at Townsville and the Commonwealth Serum Laboratories in Melbourne'. Milton Lewis, 'Introduction', in J.H.L. Cumpston, *Health and Disease in Australia*, Australian Government Publishing Service, 1989, p. 9.
26 See inside cover of *Health*, 6 (1928).
27 James Gillespie, 'The Rockefeller Foundation, the Hookworm Campaign and a National Health Policy in Australia, 1911–1930', in Roy Macleod and Donald Denoon (eds), *Health and Healing in Tropical Australia and Papua New Guinea*, James Cook University Publications, 1991, pp. 64–87.
28 Michael Roe, *Nine Australian Progressives: Vitalism in Bourgeois Social Thought 1890–1960*, University of Queensland Press, 1984.
29 Michael Roe, 'John Howard Lidgett Cumpston', *Australian Dictionary of Biography*, 1891–1939, pp. 174–6.
30 Michael Roe, *Nine Australian Progressives: Vitalism in Bourgeois Social Thought 1890–1960*, University of Queensland Press, 1984; James A. Gillespie, *The Price of Health: Australian Governments and Medical Politics 1910–1960*, Cambridge University Press, 1991, pp. 32–3. Milton Lewis, another Australian historian of Cumpston and his Department, has defined this notion of national efficiency well, as 'the purposeful application of expert or scientific knowledge to the economic, social and political spheres of national life in order to advance the power and effectiveness of the nation in a world of competitive nation states and empires'. Milton Lewis 'Introduction' to J.H.L. Cumpston, *Health and Disease in Australia*, p. 4.
31 Cited in G. Bennington, 'Postal politics and the institution of the nation', in H.K. Bhabha (ed.), *Nation and Narration*, Routledge, 1990, p. 121.
32 See Heather Bell, *Frontiers of Medicine in the Anglo-Egyptian Sudan 1899–1940*, Clarendon Press, 1999, p. 91.

33 Stern, 'Buildings, Boundaries and Blood', pp. 41–81; Bell, *Frontiers of Medicine*, ch. 6.
34 Richard Muir cited in Winichakul, *Siam Mapped*, p. 74.
35 Bynam, 'Policing Hearts of Darkness', p. 422.
36 Irving, *To Constitute a Nation*, p. 32.
37 J.S.C. Elkington, 'Quarantine in Queensland', *Australasian Medical Gazette*, 27 April 1912, p. 435.
38 Cumpston, *Health and Disease in Australia*, p. 194.
39 See, for example, Charles Mackellar, 'Federal Quarantine', *Journal of the Royal Society of New South Wales*, 17 (1883): 278–90.
40 O'Doherty, 'Federal Quarantine', p. 840. See also Elkington, 'Quarantine in Queensland', p. 435.
41 Australasian Sanitary Conference, *Report*, p. 40.
42 'Report on Quarantine', NSW Legislative Assembly, *Votes & Proceedings*, 1883: 5.
43 David Walker, *Anxious Nation: Australia and the Rise and Asia 1850–1939*, University of Queensland Press, 1999. Andrew Markus, *Fear and Hatred, Purifying Australia and California,* Hale and Iremonger, 1979
44 J.H.L. Cumpston, 'The protection of our frontiers from invasion by disease', *Australasian Medical Gazette*, 20 July 1909: 347.
45 J. Ashburton Thompson, 'Quarantine and Small-pox', *Journal of the Royal Society of New South Wales*, 21 (1887): 232.
46 Australasian Sanitary Conference, *Report*, p. 10.
47 W. Cleaver Woods, 'The Unsatisfactory Position of Vaccination in the Commonwealth', *Australasian Medical Gazette*, 20 May 1905, p. 208.
48 See, for example, Director of Public Health, Dallas, Texas to the Director of Health, 25 February 1920; Director, Antitoxin and Vaccine Laboratory, Boston MA to Minister of Health, 22 April 1925; Director-General of Health to Surgeon H.S. Cumming, US Public Health Service, 29 May 1925; Miss Mary Lee Thurman to Department of Health, 6 May 1925. National Archives of Australia (NAA) A1928 565/3.
49 Cumpston, 'The protection of our frontiers'. See also T. Borthwick, 'The Vaccination Act and its proposed amendment', *Australasian Medical Gazette*, 20 November 1900: 453.
50 Manderson argues this of the League of Nations Health Committee and its Eastern Bureau. See Lenore Manderson, 'Wireless wars in the eastern arena: epidemiological surveillance, disease prevention and the work of the Eastern Bureau of the League of Nations Health Organisation, 1925–1942', in Weindling (ed.), *International Health Organisations*, p. 116–17.
51 Cumpston, 'The protection of our frontiers', p. 437.
52 Mary Poovey, *Making a Social Body*, University of Chicago Press, 1995, chs 2, 3, 4, and 6; Erin O'Connor, *Raw Material: Producing Pathology in Victorian Culture*, Duke University Press, 2000, pp. 21–59.
53 F. Dunn, and C.R. Janes, 'Introduction, in C.R. Janes and R. Stall (eds), *Anthropology and Epidemiology*, Dordrecht: Reidl, 1986, pp. 11–12. Mervyn Susser, *Causal Thinking in the Health Sciences: Concepts and Strategies of Epidemiology*, Oxford University Press, 1972, p. 146. See also Erni's work on temporality and AIDS: 'AIDS now exists largely in time: in the definitions of the life-cycle and incubation period of the virus; in the categorization of the stages of illness for the patients; in the rate of the body's decay; in the prin-

ciple of the phases of drug development; in the "period of efficacy" of a treatment method or a drug; in the "speed" of the drug review process ...Quite literally, time becomes a field of management, something to administer'. John Nguyet Erni, *Unstable Frontiers: Technomedicine and the Cultural Politics of 'Curing' AIDS*, University of Minnesota Press, 1994, p. 70.
54 Australasian Sanitary Conference, *Report*, 1884
55 J.H.L. Cumpston, *Quarantine: Australian Maritime Quarantine and the Evolution of International Agreements Concerning Quarantine*, Government Printer, Melbourne, 1913, p.3.
56 Elkington, 'Quarantine in Queensland', p. 435.
57 Stephen Kern, *The Culture of Time and Space 1880–1918*, Harvard University Press, 1983.
58 F.F. McCallum, 'The Time Factor in Quarantine Practice', *Health*, 5 (1927): 45–6.
59 McCallum, *International Hygiene*, p. 30. See also J.H.L. Cumpston, 'Aeroplane Traffic and the Protection of Australia from Disease', *Medical Journal of Australia*, 2 September 1933, p. 326.
60 'The Trans-Pacific Flight', *Health*, 4 (1928): 97. See also McCallum, 'The Time Factor', pp. 45–51.
61 'Health Director Opposes Vaccination', *Telegraph* (Brisbane), 20 January 1939.
62 Kern, *The Culture of Time and Space*, p. 242.
63 Gillian Beer, 'The Island and the Aeroplane: the case of Virginia Woolf', in Bhabha (ed.), *Nation and Narration*, p. 265. For similar changes in the significance of international borders with the advent of air travel, see Bell, *Frontiers of Medicine*, ch. 6.
64 Martin David Dubin, 'The League of Nations Health Organisation', in Weindling (ed.), *International Health Organisations and Movements*, pp. 56–80.
65 For example, Rockefeller funds contributed to the Far Eastern Epidemiological Bureau, see Lenore Manderson, 'Wireless wars in the Eastern Arena', in Weindling (ed.) *International Health Organizations*, p. 113.
66 Cumpston, 'International Relations', in 'Report upon the Activities of the Commonwealth Department of Health'.
67 For a fuller description, see Manderson, 'Wireless wars', pp. 120–21.
68 Cumpston, 'International Relations'.
69 Manderson, 'Wireless wars', pp. 120–21.
70 Winichakul, *Siam Mapped*, pp. 1–2.
71 Australasian Sanitary Conference, *Report*, p. 15.

Chapter 6

1 David Walker, *Anxious Nation: Australia and the Rise of Asia 1850–1939*, University of Queensland Press, 1999, ch. 8.
2 Mitchell Dean, *Governmentality*, Sage, 1999, pp. 99–100.
3 See, for example, Nayan Shah, *Contagious Divides*, University of California Press, 2002; Renisa Mawani, 'Legal geographies of Aboriginal segregation in British Columbia' in Carolyn Strange and Alison Bashford (eds),

Isolation: places and practices of exclusion, Routledge, 2003 pp. 173–90.; Stefan Kuhl, *The Nazi Connection: Eugenics, American Racism, and German National Socialism*, Oxford University Press, 1994; Sheila Faith Weiss, 'The Race Hygiene Movement in Germany, 1904–1945', in Mark B. Adams (ed.), *The Wellborn Science: Eugenics in Germany, France, Brazil, and Russia*, Oxford University Press, 1990; Robert Proctor, *Racial Hygiene: Medicine under the Nazis*, Harvard University Press, 1988; Paul Weindling, *Health, Race, and German Politics between National Unification and Nazism, 1870–1945*, Cambridge University Press, 1993.

4 Anne McClintock, *Imperial Leather: Race, Gender and Sexuality in the Colonial Contest*, Routledge, 1995, pp. 207–31; Richard Dyer, *White*, Routledge, 1998.

5 A. Wallace Weihen, 'The Medical Inspection of Immigrants to Australia', *Transactions of the Australasian Medical Congress*, 1 (1911): 635.

6 Dean, *Governmentality*, pp. 138–46; Ann Laura Stoler, *Race and the Education of Desire*, Duke University Press, 1995, pp. 49–54.

7 J.W. Fortescue, 'The Influence of Climate on Race', *The Nineteenth Century*, 33 (1893): 862.

8 'Is White Australia Possible?', *Sydney Morning Herald*, 4 July 1913.

9 G.T. Searle, *Eugenics and Politics in Britain 1900–1914*, Noordhoff International Publishing, 1976; Angus McLaren, *Our Own Master Race: Eugenics in Canada, 1885–1945*, McClelland & Stewart, 1990; Daniel J. Kevles, *In the Name of Eugenics: Genetics and the Uses of Human Heredity*, Harvard University Press, 1995; Richard Soloway, *Demography and Degeneration: Eugenics and the Declining Birthrate in Twentieth Century Britain*, University of North Carolina Press, 1995; Dan Stone, *Breeding Superman: Nietzsche, Race and Eugenics in Edwardian and Interwar Britain*, Liverpool University Press, 2002.

10 David Theo Goldberg, *Racist Culture: Philosophy and the Politics of Meaning*, Blackwell, 1993. In the Australian context this has been discussed by David Dutton, *One of Us? A Century of Australian Citizenship?*, University of New South Wales Press, 2002, pp. 20–31; Warwick Anderson, *The Cultivation of Whiteness*, Melbourne University Press, 2002.

11 Stone, *Breeding Superman*, ch. 4.

12 Stephen Garton, 'Writing Eugenics: A History of Classifying Practices', in Martin Crotty, John Germov and Grant Rodwell (eds), *'A Race for a Place': Eugenics, Darwinism and Social Thought and Practice in Australia*, Faculty of Arts and Social Sciences, University of Newcastle, 2000, pp. 11–12. This idea is detailed in Russell McGregor, ' "Breed Out the Colour" or the Importance of Being White', *Australian Historical Studies*, 120 (2002): 297–301. The important developing literature on comparing genocide in Germany and Australia may be unwittingly contributing to this oversimplified conflation of eugenics with politics of racial difference, although this is not a point McGregor or Garton make. See Tony Barta, 'Discourses of genocide in Germany and Australia: a linked history', *Aboriginal History*, 25 (2001): 37–56; A. Dirk Moses, 'Coming to terms with genocidal pasts in comparative perspective: Germany and Australia', *Aboriginal History*, 25 (2001): 91–115; Paul Bartrop, 'The Composition of the Future Population:

Aboriginal Assimilation and the Jewish Immigration Restriction of the 1930s', in Crotty *et al.* (eds), *A Race for a Place*, pp. 123–32.

13 'The regulatory mechanisms of the colonial state were directed not only at the colonized, but as forcefully at "internal enemies" within the heterogenous population that comprised the category of Europeans themselves'. Stoler, *Race and the Education of Desire*, p. 96.

14 Stone, *Breeding Superman*, p. 101. See also Wolfgang Mock, 'The Function of "Race" in Imperialist Ideologies: The Example of Joseph Chamberlain', in Paul Kennedy and Anthony Nicholls (eds), *Nationalist and Racialist Movements in Britain and Germany Before 1914*, Macmillan, 1981, pp. 190–203.

15 John Torpey in *The Invention of the Passport: surveillance, citizenship and the state*, Cambridge University Press, 2000.

16 Paul Weindling (ed.), *International Health Organisations and Movements, 1918–1939*, Cambridge University Press, 1995; James A. Gillespie, 'The Rockefeller Foundation and Colonial Medicine in the Pacific, 1911–1920', in Linda Bryder and Derek Dow (eds), *New Countries, Old Medicine*, Pyramid Press, 1994, pp. 380–86.

17 C.W. Hutt, *International Hygiene*, Methuen, London, 1927.

18 One of the best summaries of the eclectic social and political problems gathered under International Hygiene is F. McCallum, *International Hygiene*, Australasian Medical Publishing Co., Sydney, 1935.

19 Hutt, *International Hygiene*, p. 115.

20 Helen Irving, *To Constitute a Nation: A Cultural History of Australia's Constitution*, Cambridge University Press, 1999, pp. 109–10; Renisa Mawani, ' "The Island of the Unclean": Race, Colonialism and "Chinese Leprosy" in British Columbia, 1891–1924', *Journal of Law, Social Justice and Global Development*, (2003) http.//elj.warwick.ac.uk/global/

21 Hutt, *International Hygiene*.

22 R.A. Huttenback, *Racism and Empire: white settlers and colored immigrants in the self-governing colonies, 1830–1910*, Cornell University Press, 1976; Stainslaw Andracki, *Immigration of Orientals into Canada, with Special reference to the Chinese*, Arno Press, 1978; A.T. Yarwood and M.J. Knowling, *Race Relations in Australia: A History*, Methuen Australia, 1982, p. 177. See also W. Peter Ward, *White Canada Forever: Popular Attitudes and Public Policy Towards Orientals in British Columbia*, McGill-Queens University Press, 1990.

23 Torpey, *The Invention of the Passport*, p. 97; Hutt, *International Hygiene*, p. 116; Gregory, *Human Migration and the Future*, p. 70.

24 He thought this an entirely misplaced opinion. See John Pope Hennessy, 'The Chinese in Australia', *The Nineteenth Century*, 23 (1888): 618.

25 Persia Crawford Campbell, *Chinese Coolie Emigration to Countries within the British Empire*, P.S. King, London, 1923.

26 A.T. Yarwood, 'The Overseas Indians: A Problem in Indian and Imperial Politics at the end of World War One', *Australian Journal of Politics and History*, 15 (1968): 204–18; R.A. Huttenback, 'No Strangers Within the Gates: Attitudes and Policies towards the non-white residents of the British Empire of Settlement', *Journal of Imperial and Commonwealth History*, 1 (1972–3): 271–302.

27 For full detail of this see S. Brawley, *The White Peril: Foreign Relations and Asian Immigration to Australasia and North America 1901–1978*, University of New South Wales Press, 1995.
28 H.L. Wilkinson, *The World's Population Problems and a White Australia*, P.S. King, London, 1930, pp. 215–17.
29 J.W. Gregory, *Human Migration and the Future: A study of the causes, effects and control of emigration*, Seeley, Service & Co., London, 1928, pp. 49, 70.
30 Cited in Wilkinson, *The World's Population Problems and a White Australia*, p. 167; Carolyn Strange and Tina Loo, *Making Good: Law and Moral Regulation in Canada, 1867–1939*, University of Toronto Press, 1997, pp. 76, 120; see also McLaren, *Our Own Master Race*, ch. 3.
31 Wilkinson, *The World's Population Problems and a White Australia*, p. 169.
32 Section 3 a) *Immigration Restriction Act*, 1901.(cth)
33 Yarwood, *Asian Migration to Australia*, p. 15, p. 27. Huttenback, *Racism and Empire*, ch. 3.
34 Wilkinson, *The World's Population Policy and a White Australia*, p. 169.
35 See, for example, Sir Macfarlane Burnet's essays 'Biology and Medicine', *Eugenics Review*, 49 (1957): 127–35; and 'Migration and Race Mixture from the Genetic Angle', *Eugenics Review*, 51 (1959): 93–7. See also George A. Gellert, 'International Migration and Control of Communicable Diseases', *Social Science and Medicine*, 37 (1993): 1489–99; and Jay K. Varma, 'Eugenics and Immigration Restriction: lessons for tomorrow', *Journal of the American Medical Association*, 275 (1996): 734.
36 W.E. Agar, 'Some Eugenic Aspects of Australian Population Problem', in P.D. Phillips and G.L. Wood (eds), *The People of Australia*, Macmillan, Melbourne, 1928, p. 130.
37 'Memorandum on Alien Immigration', Eugenics Society Papers, Wellcome Library, SA/EUG/D103, n.d.
38 Kenneth M. Ludmerer, 'Genetics, Eugenics and the Immigration Restriction Act of 1924', *Bulletin of the History of Medicine*, 46 (1972): 59–81; Hutt, *International Hygiene*, pp. 115–21; McLaren, *Our Own Master Race*, ch. 3; see also Mathew Thomson, *The Problem of Mental Deficiency: Eugenics, Democracy and Social Policy in Britain, c. 1870–1959*, Clarendon Press, 1998. On mental hygiene in Australia, see Stephen Garton, 'Sound Minds and Healthy Bodies: Re-considering eugenics in Australia', *Australian Historical Studies*, 26 (1994): 163–81; Ross L. Jones, 'The Master Potter and the Rejected Pots: Eugenic Legislation in Victoria 1918–1939', *Australian Historical Studies*, 113 (1999): 319–42. See also Bernard Harris, 'Anti-Alienism, Health and Social Reform in Late-Victorian and Edwardian Britain', *Patterns of Prejudice*, 31 (1997): 3–34.
39 For background on the Society, see Soloway, *Demography and Degeneration*, pp. 31–37.
40 'Memorandum on Alien Immigration' SA/EUG/D103. For further discussion of eugenics and immigration into Britain, see Stone, *Breeding Superman*, pp. 94–114.
41 This set of letters are in the Eugenics Society Papers, SA/EUG/D103 and 105. Most are undated except for one with the year 1926 and a letter from Mr Bruce's private Secretary to Leonard Darwin acknowledging receipt of materials from the Eugenics Education Society, 27 November 1926.

42 Yarwood and Knowling, *Race Relations in Australia*, pp. 183–5.
43 This was carried on the nationalist journal *The Bulletin* well into the twentieth century.
44 Sir Henry Parkes, 16 May 1888, quoted in A.T. Yarwood (ed.), *Attitudes to Non-European Immigration*, Cassell, 1968, p. 94.
45 Commonwealth *Parliamentary Debates*, 6 September 1901, vol. 4, p. 4633.
46 Commonwealth *Parliamentary Debates*, 12 September 1901, vol. 4, p. 4845.
47 Huttenback examines some contestations in *Racism and Empire*, ch. 2.
48 This is a large literature but for overview works, see A.T. Yarwood, *Asian Migration to Australia; the Background to Exclusion, 1896–1923*, Melbourne University Press, 1964; Yarwood and Knowling, *Race Relations in Australia*; Ann Curthoys and Andrew Markus (eds), *Who are our Enemies? Racism and the Working Class in Australia*, Hale & Iremonger, 1978; Andrew Markus, *Fear and Hatred: Purifying Australia and California 1850–1901*, Hale & Iremonger, 1979; and Andrew Markus, *Australian Race Relations*, Allen & Unwin, 1994; R. Evans et al., *Exclusion, Exploitation and Extermination*, Australia and NZ Book Co., 1975; A.-M. Jordens, *Redefining Australians: immigration, citizenship and national identity*, Hale & Iremonger, 1995; Irving, *To Constitute a Nation*, ch. 6.
49 F.M. Cutlack, 'The White Australia Question', *The Empire Review*, 17 (1909): 286.
50 Catriona Elder, 'Dreams and Nightmares of a White Australia', PhD thesis, Australian National University, 1999; Russell McGregor, *Imagined Destinies*, Melbourne University Press, 1994.
51 McCallum, health bureaucrat and author of *International Hygiene*, wrote, for example, that 'the national alloy is very precious metal. Nothing in Australian health history leads one to oppose the ideal and policy of a White Australia. Unlike those of other countries, the aborigines have played no part in the epidemiological history of the white community in Australia. The tragedy of their decline is another story.' F. McCallum. 'Bionomics of Australian History', *Health*, 4 (1926): 50.
52 Sheila Fitzpatrick, *Red Tape, Gold Scissors: The Story of Sydney's Chinese*, State Library of New South Wales Press, 1996, esp. ch. 1.
53 Peter Corris, ' "White Australia" in Action: the repatriation of Pacific Islanders from Queensland', *Historical Studies*, 15 (1972): 237–50.
53 Section 3d. Immigration Restriction Act, 1901 (Cth).
54 Irving, *To Constitute a Nation*, p. 100.
55 Ibid., p. 101.
56 See Huttenback, *Racism and Empire*, pp. 85–6; Greg Watters, 'The S.S. Ocean: Dealing with Boat people in the 1880s', *Australian Historical Studies*, 120 (2002): 331–43.
57 Yarwood, *Asian Migration to Australia*, p. 5.
58 Campbell, *Chinese Coolie Emigration*, p. 57.
59 Cited in Irving, *To Constitute a Nation*, p. 115.
60 Although alongside the Immigration Restriction Act, the nation/race specific Pacific Island Labourers Act instituted the gradual deportation of indentured Islander labourers over several years, recognising the particular needs of the Queensland industry and its reliance on the

labour of Pacific Islanders. See Corris, ' "White Australia" in Action', 237–50.
61 Section 3d. Immigration Restriction Act, 1901 (Cth).
62 J.H.L. Cumpston, 'Cleanliness'. Unpublished typescript, Cumpston Papers, National Library Canberra, MS 613 Box 7 (1).
63 See NSW Premier, Quarantine Service Minute Paper, 25 February 1913, Department of Health Quarantine Papers, 1907–1914, Department of Health and Community Services Library, Canberra.
64 In the 1925 Regulations for the Immigration Act, for example, the Health Reports required by the medical officer or master of a ship were rewritten to correlate with the requirements of the Quarantine Act. See Immigration Regulations 1926, no. 185, p. 868.
65 Cumpston to the Secretary, Department of Home and Territories, 26 January 1921, NAA AI/15 1921/12036.
66 See letters and telegrams from 1921 detailing these procedures collected in 'Alteration of System of issuing Medical Certificates under Immigration Act at Darwin, Thursday Island', NAA A1/15 1921/12036.
67 War on Foreign Germs (1 Feb. 1933), Album of Newsclippings, 1913–45, Cumpston Papers, MS 613 Box 8 (iv).
68 While the Commonwealth permitted the entry of Europeans as well as British, the vast majority of migrants in the first half of the twentieth century were British. See Michael Roe, *Australia, Britain and Migration, 1915–1940*, Cambridge University Press, 1995. For detailed analysis of the debate over European immigration in the period, see Dutton, *One of Us?*, pp. 44–60.
69 Australasian Sanitary Conference, *Report and Minutes of Proceedings*, Government Printer, Sydney, 1884, p. 53.
70 Immigration Regulations, 1913, no. 307, p. 1058.
71 Section 3, Immigration Act, 1912.
72 Section 4, 3A, Immigration Act, 1912.
73 Weihen, 'The Medical Inspection of Immigrants to Australia', pp. 637–9. See also 'Prohibited Immigrants – by One of Them', undated typescript in Cumpston Papers, MS 613 Box 7, p. 1.
74 Quarantine Regulations, 1915, section 56, p. 515.
75 Quarantine Regulations, 1917.
76 Anthea Hyslop, 'Insidious Immigrant: Spanish Influenza and Border Quarantine in Australia, 1919', in S. Barry and B. Reid (eds), *Migration to Mining*, Northern Territory University Press, 1997; Anthea Hyslop, 'Old Ways, New Means: Fighting Spanish Influenza in Australia, 1918–1919', in Linda Bryder and Derek A. Dow (eds), *New Countries, Old Medicine*, Pyramid Press, 1995, pp. 46–53.
77 For a history of the Association, see Stefania Siedlecky and Diana Wyndham, *Populate and perish: Australian women's fight for birth control*, Allen & Unwin, 1990.
78 Marie Carmichael Stopes, *Wise Parenthood*, G.P. Putnam, London, 1918.
79 This is evident in any of the Annual Reports of the Racial Hygiene Association. See also the newspaper clippings in the Association's papers, Newspaper cuttings 1927–35, Family Planning Association Records, ML MSS 3838, Mitchell Library, Sydney.

80 In this period, 'Native' often referred to British-whites who were native born, that is born in Australia. For the idea of the Australian Native and the Australian Natives Association, see Irving, *To Constitute a Nation*, ch. 7.
81 'A Report on Immigration (as affecting Racial Values and Public Health in NSW)' in Report on Immigration with regard to Racial Health, 1928, NAA A458 2154/1.
82 Advisory Committee of the Racial Hygiene Association of NSW to Stanley Bruce, 13 January 1928, in Report on Immigration (as affecting Racial Values).
83 'A Report on Immigration (as affecting Racial Values)', pp. 16–17.
84 David Walker discusses this at length in *Anxious Nation*, pp. 113–26.
85 Charles Edward Woodruff, *Expansion of Races*, Rebman, London, 1909.
86 Walker, *Anxious Nation*, p. 126.
87 'A Report on Immigration (as affecting Racial Values)', p. 7.
88 The negotiations between British governments and Australian governments over the Empire Settlement Program, the Big Brother Movement and more are examined by Michael Roe in *Australia, Britain and Migration*. See also Geoffrey Sherington, ' "A Better Class of Boy?" The Big Brother Movement, Youth Migration and Citizenship of Empire', *Australian Historical Studies*, 120 (2002): 267–85.
89 Fleetwood Chidell, *Australia – White or Yellow?*, Heinemann, London, 1926, ch. 2.
90 Gregory, *Human Migration and the Future*, p. 149.
91 John Farley, *Bilharzia: A History of Imperial Tropical Medicine*, Cambridge University Press, 1991, p. 3. Douglas M. Haynes, *Imperial Medicine: Patrick Manson and the Conquest of Tropical Disease*, University of Pennsylvania Press, 2001; Michael Worboys, 'Manson, Ross and Colonial Medical Policy: Tropical Medicine in London and Liverpool, 1899–1914', in Roy Macleod and Milton Lewis (eds), *Disease, Medicine and Empire*, Routledge, 1988, pp. 26–7.
92 See, for example, Kenneth F. Kiple and Kriemhild Conee Ornelas, 'Race, War and Tropical Medicine in the Eighteenth-Century Caribbean', in David Arnold (ed.), *Warm Climates and Western Medicine*, Rodopi, pp. 65–79; Trevor Burnard', ' "The Countrie Continues Sicklie": White Mortality in Jamaica, 1655–1780', *Social History of Medicine*, 12 (1999): 45–72; Mark Harrison, *Climates and Constitutions: Health, Race, Environment and British Imperialism in India 1600–1850*, Oxford University Press, 1999.
93 Warwick Anderson, 'Climates of Opinion: Acclimatization in Nineteenth-Century France and England', *Victorian Studies*, 35 (1992): 1–24; David N. Livingstone (ed.), *The geographical tradition: episodes in the history of a contested enterprise*, Blackwell, 1993; Lenore Manderson, *Sickness and the State: Health and Illness in Colonial Malaya*, Cambridge University Press, 1996; David Arnold, 'Introduction; Tropical Medicine before Manson', in Arnold (ed.), *Warm Climates and Western Medicine*, pp. 5–9; David N. Livingstone, 'Tropical climate and moral hygiene: the anatomy of a Victorian debate', *British Journal of the History of Science*, 32 (1999): 93–110.

94 Alison Bashford, ' "Is White Australia Possible?" race, colonialism and tropical medicine', *Ethnic and Racial Studies*, 23 (2000): 248–71.
95 This has been discussed extensively in Australian historiography. See Michael Roe, *Nine Australian Progressives: Vitalism in Bourgeois Social Thought*, University of Queensland Press, 1984; Helen R. Woolcock, ' "Our Salubrious Climate": attitudes to health in colonial Queensland', in Macleod and Lewis (eds), *Disease, Medicine and Empire*, pp. 176–93; Lorraine Harloe, 'Anton Breinl and the Australian Institute of Tropical Medicine', in Roy Macleod and Donald Denoon (eds), *Health and Healing in Tropical Australia and Papua New Guinea*, 1991; A.T. Yarwood, 'Sir Raphael Cilento and The *White Man in the Tropics*, in Ibid; James A. Gillespie, *The Price of Health: Australian Governments and Medical Politics 1910–1960*, Cambridge University Press, 1991, pp. 41–3; David Walker, 'Climate, Civilization and Character in Australia, 1880–1940', *Australian Cultural History*, 16 (1997/98): 77–95; Walker, *Anxious Nation*, chs 11 and 12; Anderson, *The Cultivation of Whiteness*, chs 3–6.
96 See also Yarwood, 'Sir Raphael Cilento and *The White Man in the Tropics*', p. 51 ff.; and Harloe, 'Anton Breinl and the Australian Institute of Tropical Medicine', p. 34.
97 J.H.L. Cumpston, 'Report upon the Activities of the Commonwealth Department of Health, 1930, section 3.
98 J.H.L. Cumpston, *The Health of the People: a study in federalism*, Roebuck, Canberra, 1978, p. 49.
99 Anton Brienl, 'The Object and Scope of Tropical Medicine in Australia', *Australasian Medical Congress Transactions*, 1911, pp. 524–5. Warwick Anderson details the establishment and work of the Institute in *The Cultivation of Whiteness*, ch.4.
100 *Tropical Australia: Report of the Discussion at the Australasian Medical Congress, 1920*, Government Printer, Melbourne, 1921.
101 Walker, 'Climate, Civilization and Character in Australia', pp. 86–90; Anderson, *The Cultivation of Whiteness*, pp. 165–72.
102 A. Breinl and W.J. Young, 'Tropical Australia and its Settlement', in Australian Institute of Tropical Medicine, *Collected Papers*, 3 (1922): 1–24.
103 J.W. Barrett, 'Can Tropical Australia be Peopled by a White Race?', *The Margin*, 1 (1925): 29.
104 See *Tropical Australia*, p. 13.
105 These studies are detailed in *Tropical Australia*. See also The Australian Institute of Tropical Medicine, *Collected Papers*, no. 2, Townsville, 1917.
106 See, for example, W.J. Young, A. Breinl, J.J. Harris and W.Z. Osborne, 'Effect of Exercise and Humid Heat upon Pulse Rate, Blood Pressure, Body Temperature, and Blood Concentration', Australian Institute of Tropical Medicine, *Collected Papers*, 3 (1922): 111–25.
107 See, for example, W.J. Young, 'The Metabolism of White Races Living in the Tropics' *Annals of Tropical Medicine and Parasitology*, 9 (1915): 91–108; W.J. Young, 'Observations upon the Body Temperature of Europeans Living in the Tropics', *Journal of Physiology*, 49 (1915): 222–32.
108 The contents list runs thus: Blood Sugar, Non-Protein Nitrogen of the Blood, Phosphorus of the Blood, Lipoid Constituents of the Blood, Water

Regulation, Acid-Base Equilibrium, Basal Metabolism, Dietary Experiments, Urinary Analysis, Growth of Children, Growth of Hair and Nails, Experiments on Growth of Rats, Measurements of Cooling Power, Red and White Blood Corpuscles, Reaction Time to Stimuli. E.S. Sundstroem, *Contributions to Tropical Physiology: with special reference to the adaption of the white man to the climate of North Queensland*, University of California Publications in Physiology, vol. 6, 1926.
109   A. Grenfell, 'The White Man in the Tropics', *Medical Journal of Australia*, 26 January 1935: 106.
110   Walker, *Anxious Nation*, ch. 12; Anderson discusses the geographers' debates in detail in *The Cultivation of Whiteness*, pp. 164–74.
111   Fleetwood Chidell, *Australia White or Yellow*, cited in Wilkinson, *The World's Population Problems and a White Australia*, p. 201. See also Walker, *Anxious Nation*, p. 125.
112   Cutlack, 'The White Australia Question', p. 293.

Chapter 7

1   For studies which deftly link sexology, feminism, health and eugenics, see Roy Porter and Lesley Hall, *The facts of life: the creation of sexual knowledge in Britain, 1650–1950*, Yale University Press, 1995; Lesley Hall, 'Feminist Reconfigurations of heterosexuality in the 1920s', in Lucy Bland and Laura Doan (eds), *Sexology in culture: labelling bodies and desires*, University of Chicago Press, 1998, pp. 135–49; Carolyn Burdett, 'The Hidden Romance of Sexual Science; Eugenics, the Nation and the Making of Modern Feminism', ibid., pp. 45–59.
2   Links between public health and eugenics have been explored in the British context in Dorothy Porter, ' "Enemies of the Race": Biologism, Environmentalism and Public Health in Edwardian England', *Victorian Studies*, 34 (1991): 159–78. See also Dorothy Porter, *Health, Civilization and the State: a history of public health from ancient to modern times*, Routledge, 1999; Peter Weingart, 'The Thin Line Between Eugenics and Preventive Medicine', in Nobert Finzsch and Dietmar Schirmer (eds), *Identity and Intolerance: Nationalism, Race, and Xenophobia in Germany and the United States*, Cambridge University Press, 1998, pp. 397–412.
3   For summaries and extensions of the historiography of eugenics in Australia, see Stephen Garton, 'Sound Minds and Healthy Bodies: Re-considering Eugenics in Australia, 1914–1940', *Australian Historical Studies*, 26 (1994): 163–81; Ross L. Jones, 'The Master Potter and the Rejected Pots: Eugenic Legislation in Victoria 1918–1939', *Historical Studies*, 113 (1999): 319–42; Martin Crotty, John Germov and Grant Rodwell (eds), *'A Race for a Place': Eugenics, Darwinism and Social Thought and Practice in Australia*, Faculty of Arts and Social Sciences, University of Newcastle, 2000.
4   Ann Laura Stoler, *Race and the Education of Desire*, Duke University Press, 1996, pp. 48, 69.
5   D. Peukert, 'The genesis of the "Final Solution" from the sprit of science', in T. Childers and J. Caplan (eds), *Re-evaluating the Third Reich*, Holmes & Meier, 1993, pp. 234–52; Atina Grossman, *Reforming Sex: The German*

Movement for Birth Control and Abortion Reform, 1920–1950, Oxford University Press, 1995; Paul Weindling, *Health, Race, and German Politics between National Unification and Nazism, 1870–1945*, Cambridge University Press, 1993; Robert Proctor, *Racial Hygiene: Medicine under the Nazis*, Harvard University Press, 1988.

6   Michel Foucault, 'The politics of health in the eighteenth century', in P. Rabinow (ed.), *The Foucault Reader*, Penguin, 1984, p. 277.
7   For a summary of social theorists' interest in the period and the problem, see Mitchell Dean, *Governmentality*, Sage, 1999, ch. 7.
8   Stoler, *Race and the Education of Desire*, 1996, p. 20.
9   Alan Petersen and Deborah Lupton, *The New Public Health: health and self in the age of risk*, Allen & Unwin, 1996.
10 Garton argues that the significance of eugenics lies largely in its dispersal of the classificatory impulse. See Stephen Garton, 'Writing Eugenics: a history of classifying practices', in Crotty *et al.* (eds), *'A Race for a Place'*, pp. 9–18.
11 Stephen Garton, 'Policing the Dangerous Lunatic: Lunacy Incarceration in New South Wales, 1843–1914', in Mark Finnane (ed.), *Policing in Australia: Historical Perspectives*, University of New South Wales Press, 1987, pp. 74–87.
12 Philippa Levine, *Prostitution, Race and Politics: policing venereal disease in the British Empire*, Routledge, 2003.
13 For India, see Douglas Peers, 'Soldiers, Surgeons and the Campaigns to Combat Sexually Transmitted Diseases in Colonial India, 1805–1860', *Medical History*, 42 (1998): 137–60; for Australia, see Milton Lewis, *Thorns on the Rose: The History of Sexually Transmitted Diseases in Australia in International Perspective*, Australian Government Publishing Service, 1998, p. 94; Kay Saunders and Helen Taylor, ' "To Combat the Plague": The Construction of Moral Alarm and State Intervention in Queensland During World War II', *Hecate*, 14 (1988): 5–30.
14 This is detailed in Saunders and Taylor, '"To Combat the Plague"', pp. 5–30. See also Mary Murnane and Kay Daniels, 'Prostitutes as "Purveyors of Disease": Venereal Disease Legislation in Tasmania, 1868–1945', *Hecate*, 5 (1979): 5–21.
15 See also the instances in Carolyn Strange and Alison Bashford (eds), *Isolation: places and practices of exclusion*, Routledge, 2003.
16 Dr Arthur Adams cited in Mary Ann Jebb, 'The Lock Hospitals Experiment: Europeans, Aborigines and Venereal Disease', *European-Aboriginal Relations in Western Australian History*, 8 (1984): 74.
17 Ibid., pp. 68–87; See also Lewis, *Thorns on the Rose*, pp. 374–79.
18 Claudia Thame, 'Health and the State in Australia', PhD thesis, Australian National University, 1974, p. 118.
19 Section 3.1 Prisoners' Detention Act, 1908 (NSW).
20 Progress Report from the Select Committee on the Prevalence of Venereal Diseases, New South Wales Legislative Assembly, *Votes and Proceedings*, 1915 (Hereafter Select Committee on Venereal Diseases).
21 For detail on venereal disease in New South Wales in the period, see Greg Ussher, 'The 'medical gaze' and the 'watchful eye': the prevention, treatment and epidemiology of venereal diseases in NSW 1900–1925, PhD thesis, University of Sydney, forthcoming. My thanks to Greg Ussher for discussion on these points.

22 Evidence about the Liverpool Camp in Sydney was sought by the Select Committee on Venereal Diseases, 1915.
23 See Judith Smart, 'Sex, the State and the "Scarlet Scourge": gender, citizenship and venereal disease regulation in Australia during the Great War', *Women's History Review*, 7 (1998): 5–36; Milton Lewis also details Australian responses to venereal disease management in wartime. See *Thorns on the Rose*, pp. 153–70, 246–61.
24 Bernadette McSherry, ' "Dangerousness" and public health', *Alternative Law Journal*, 57 (1998): 276–80; Alison Bashford and Carolyn Strange, 'Asylum Seekers and National Histories of Detention', *Australian Journal of Politics and History*, 48 (2002): 509–27.
25 'A Threat of Compulsion!', *Health and Empire*, 12 (1937): 1.
26 'Social Hygiene and General Physical Fitness', *Health and Empire*, 13 (1938): 200.
27 'A Threat of Compulsion!', p. 2.
28 'The Imperial Aspects of Social Hygiene', *Health and Empire*, 1 (1926): 1.
29 *Health and Empire*, 2 (1927): iv.
30 'A Threat of Compulsion!', p. 2.
31 Racial Hygiene Association, Minute Books, 27 April 1926, Family Planning Association Records, Mitchell Library, Sydney (ML) MSS 3838.
32 For example, the Australasian White Cross League, the Workers' Educational Association, the Australian Association for Fighting Venereal Disease, the University of Sydney Society for Combatting Venereal Diseases. See Thame, Health and Disease in Australia', p. 135; Lewis, *Thorns on the Rose*, pp. 174–80, 187–94.
33 Paul Weindling. 'Public Health in Germany', in Dorothy Porter (ed.), *The History of Public Health and the Modern State*, Rodopi, 1994, p. 122.
34 See Alisa Klaus, 'Depopulation and Race Suicide: Maternalism and Pronatalist Ideologies in France and the United States', in Seth Koven and Sonya Michel (eds), *Mothers of a New World: Maternalist Politics and the Origins of Welfare States*, Routledge, 1993, pp. 188–212; Richard Soloway, *Demography and Degeneration*, University of North Carolina Press, 1995; William H. Schneider, *Quality and Quantity: The Quest for Biological Regeneration in Twentieth Century France*, Cambridge University Press, 1990; Rosemary Pringle, 'Octavius Beale and the Ideology of the Birth-Rate', *Refractory Girl*, 3 (1973): 19–27.
35 James Marchant, *Birth-Rate and Empire*, Williams and Norgate, London, 1917; Anna Davin, 'Imperialism and Motherhood', in Frederick Cooper and Ann Laura Stoler (eds), *Tensions of Empire*, University of California Press, 1997.
36 Lynette Finch, *The Classing Gaze: Sexuality, Class and Surveillance*, Allen & Unwin, 1993, ch. 6; Alison Mackinnon, *Love and Freedom? Professional Women and the Reshaping of Personal Life*, Cambridge University Press, 1997, chs 2, 3.
37 See Marilyn Lake, *Getting Equal: The history of Australian feminism*, Allen & Unwin, 1999, chs 2, 3; Patricia Grimshaw, Marilyn Lake, Ann McGrath and Marian Quartly, *Creating a Nation*, McPhee Gribble, 1994, chs 8, 9.
38 Kerreen M. Reiger, *The disenchantment of the home: Modernizing the Australian family*, Oxford University Press, 1985; Alison Bashford, 'Separatist Health: Meanings of Women's Hospitals in England and Australia, c. 1870–1930', in Lilian R. Furst (ed.), *Climbing a Long Hill: Women Healers and Physicians*, University Press of Kentucky, 1997,

pp. 198–220; Philippa Mein Smith, *Mothers and King Baby: Infant Survival and Welfare in an Imperial World: Australia 1880–1950*, Macmillan, 1997; Philippa Mein Smith, 'Maternity and Eugenics', in Crotty *et al.* (eds), '*A Race for a Place*', pp. 141–56.
39 For comparative studies, see Gisela Bock and Pat Thane (eds), *Maternity and Gender Policies: Women and the Rise of the European Welfare States, 1880s–1950s*, Routledge, 1991.
40 Editorial, *Health and Empire*, 12 (1937): 265.
41 Minutes of the Propaganda Committee 28 November 1928, British Social Hygiene Council Records, Wellcome Library, SA/BSH C.3.
42 Aims and Objectives in *Health and Empire*, 7 (1932).
43 'Health Propaganda' n.d., Eugenics Society Papers, Wellcome Library, SA/EUG/G 29.
44 Mrs A.B. Piddington, 'Making Australia Healthy for Unborn Generations', *Smiths Weekly*, 23 January 1932, newsclipping in Family Planning Association Records, ML MSS 3838.
45 For other enterprises which preferred advice and education in Australia see Ann Curthoys, 'Eugenics, Feminism and Birth Control: The Case of Maion Piddington', *Hecate*, 15 (1989): 73–89.
46 Piddington, 'Making Australia Healthy'.
47 Dr Arthur, 'Certificate of Health Prior to Marriage and Sterilisation', Racial Hygiene Association One Day Conference in connection with Health Week, 1931, p. 7. Typescript in ML.
48 Lionel Lewis to W.S.S. Hoodson, Secretary, the Eugenics Education Society, 16 June 1926, Eugenics Society Papers, SA/EUG/E.3.
49 See Jones, 'The Master Potter and the Rejected Pots', pp. 319–42; Gunnar Broberg and Nils Roll-Hansen (eds), *Eugenics and the Welfare State: Sterilization Policy in Denmark, Sweden, Norway and Finland*, Michigan State University Press, 1996.
50 Dr Blacker to Mrs Angela Booth, 23 September 1938, Eugenics Society Papers, SA/EUG/E.3/1; Jones, 'The Master Potter'.
51 Eugenics Society of Victoria, 'Statement of Principles Suggested for Acceptance by the Society', November 1947.
52 Jan Kociumbas, 'Reflecting on "the Century of the Child": Child Study and the School Medical Service in New South Wales', in Crotty *et al.* (eds), '*A Race for a Place*', pp. 221–8; Lewis, *Thorns on the Rose*, pp. 172–4.
53 Rev. D.P. McDonald, 'Health and Education', Racial Hygiene Association One Day Conference in connection with Health Week, 1931.
54 Klaus Theweleit, *Male Fantasies*, trans. Stephen Conway, University of Minnesota Press, 1989.
55 Rev. H.N. Baker, 'The Wider Implications of the Policy of Sterilisation', in Racial Hygiene Association One Day Conference in connection with Health Week, 1931, p. 16.
56 Matthew Thomson, *The Problem of Mental Deficiency: Eugenics, Democracy and Social Policy in Britain, c. 1870–1959*, Clarendon Press, pp. 110–48; Stephen Garton, *Medicine and Madness: A Social History of Insanity in New South Wales 1880–1940*, University of New South Wales Press, 1988, p. 60.
57 Baker, 'The Wider Implications of the Policy of Sterilisation', p. 14; see also Garton, *Medicine and Madness*, pp. 60–62.

58 'In many of these cases freedom with sterilization is more humane than confinement in an institution'. Professor W.E. Agar, *Eugenics and the Future of the Australian Population*, Brown, Prior, Anderson, Melbourne, 1939, p. 7.
59 Thomson, *The Problem of Mental Deficiency*, pp. 198–205.
60 Dean, *Governmentality*, p. 140.
61 The President of the Eugenics Society of Victoria wrote: 'The governments of many countries have taken the view that their responsibility to posterity involves legislation to provide for the sterilization of persons likely to transmit mental disabilities'. He pointed in 1939 to various states in the United States, to Germany, Denmark, Norway, Sweden, and Alberta and British Columbia in Canada. See Agar, *Eugenics and the Future of the Australian Population*, 1939, p. 7.
62 Other have written about this, and more research into practice as well as social and political debate needs to be undertaken. Thomson, *The Problem of Mental Deficiency*, pp. 202–4; Jones, 'The Master Potter', see also Daniel Pick, *The Faces of Degeneration: A European Disorder, c.1848–c.1918*, Cambridge University Press, 1989.
63 Eugenics Education Society, *Those Who Come After: A Word on Social Superiority*, n.d. Eugenics Society Papers, SA/EUG/J.17.
64 H.L. Wilkinson, *The World's Population Problems and a White Australia*, P.S. King & Sons, London, 1930.
65 Ibid.
66 A.W. Hayes, *Future Generations: Woman the Future Ruler of this Earth*, Sydney, privately printed, 1915, p. 7 (ML).
67 Hayes, *Future Generations*, p. 12.
68 Mr Creswell O'Reilly speaking at the Racial Hygiene Association One Day Conference, 1931, p. 17.
69 Eugenics Education Society (Victoria) *Annual Report*, 1939 in Eugenics Society Papers, SA/EUG/E.3/1.
70 Eugenics Education Society, *Those Who Come After: A Word on Social Superiority*, n.d. Eugenics Society Papers, SA/EUG/J.17.
71 Cited in Eugenics Education Society of NSW Report to Eugenics Education Society, 9 November 1921, SA/EUG/E.2.
72 See Stoler, *Race and the Education of Desire*, p. 52.
73 Dr Arthur, 'Certification of Health Prior to Marriage and Sterilisation', p. 8. Another contributor to the Conference said, 'One element in this subject of heredity is to discover how bad strains in human propagation may be eliminated. There are many such bad strains, moral and physical and mental, which tend to perpetuate themselves through succeeding generations'. Baker, 'The Wider Implications of the Policy of Sterilisation', p. 14.
74 See Jean-Paul Gaudillière and Ilana Löwy, 'Horizontal and Vertical Transmission of Disease', in Gaudillière and Löwy (eds), *Heredity and Infection: The History of Disease Transmission*, Routledge, 2001, pp. 1–18.
75 Nikolas Rose, *Powers of Freedom: reframing political thought*, Cambridge University Press, 1999.
76 Eugenics Education Society, *Teaching in School, Training Colleges and Colleges: from the point of view of the eugenist*, n.d. p. 1, Eugenics Society Papers, SA/EUG/J.17.

Conclusion

1 Bernadette McSherry, ' "Dangerousness" and public health', *Alternative Law Journal*, 57 (1998): 276–80; Richard Coker, *From Chaos to Coercion: Detention and the Control of Tuberculosis*, St Martin's Press, 2000; Richard Coker, 'Civil Liberties and Public Good: Detention of Tuberculous Patients and the Public Health Act 1984', *Medical History*, 45 (2001): 339–56.
2 J.H.L. Cumpston, 'Cleanliness', typescript in Cumpston Papers, National Library of Australia, MS 613 Box 7, p. 3.
3 Ibid., p. 8.
4 Ibid., p. 4.
5 Ibid., p. 13.
6 Ibid., p. 13.
7 Ibid., p. 14.
8 J.H.L. Cumpston, Report upon The Activities of the Commonwealth Department of Health from 1909 to 1930, typescript, Department of Health Library, Canberra, 1930, unpaginated (section 1).
9 Nikolas Rose, 'Governing "advanced" liberal democracies', in Andrew Barry, Thomas Osborne and Nikolas Rose (eds), *Foucault and Political Reason*, University of Chicago Press, 1996, pp. 45–6.

Select Bibliography

A. Primary sources

A.1 Archival sources

New South Wales State Archives

Medical officer reports and returns, applications for appointments as vaccinators 1869–1874 4/790.1.
Board of Health Records 1881–1896 5/2913.
Board of Health minutes 1882–85 5/5837.
Quarantine Books 5/5853–4.
North Head Quarantine Station 1909–1930 5/5396.
Chief Secretary's Department, Smallpox Files, 1913–15, 5/5290.
Waterfall Sanatorium, Case Histories, 1909, Colonial Secretary's Special Bundle, X648.

Mitchell Library, Sydney

Newspaper cuttings on Tuberculosis, 1901–17, Folio 616.2/N.
Family Planning Association Records, MSS 3838.

Wellcome Library for the History and Understanding of Medicine, London

Sir Leonard Rogers Papers, PP/ROG.
Eugenics Society Papers, SA/EUG.
British Social Hygiene Council Records, SA/BSH.

National Archives of Australia, Canberra

Leprosy in the Commonwealth, A1 1908/4507.
Reports on Tuberculosis, A431 1949/422.
Report on Immigration with regard to Racial Health, 1928, A458 2154/1.

National Archives of Australia, Melbourne

Immigration Act 1901–1925 Deportation for Health Reasons, Series B13.
Medical reports on, and records of restricted passengers and crew members, Series B13.

Queensland State Archives, Brisbane

Leprosy files, Queensland Home Secretary's Office, COL 266, COL 322, COL 323, COL 324.

Fryer Library, Brisbane

Sir Raphael Cilento Collection.

National Library of Australia, Canberra
Cumpston Papers, MS 613.

Department of Health Library, Canberra
Department of Health Quarantine Papers, 1907–1914.
J.H.L. Cumpston, Report upon the Activities of the Commonwealth. Department of Health from 1909 to 1930.

A.2 Statutes

An Act for the Prevention of the Disease called the Cholera, 2 and 3, William IV, c. 10, 1832.
Compulsory Vaccination Act 1853 (Tasmania).
Act to extend and make compulsory the practice of Vaccination, 1853 (South Australia).
Act to Make Compulsory the Practice of Vaccination, 1854 (Victoria).
An Ordinance to Make Compulsory the Practice of Vaccination, 1860 (Western Australia).
Infectious Disease Supervision Act, 1881 (NSW).
Immigration Restriction Act, 1901 (Cth).
Quarantine Act, 1908 (Cth)
Prisoners' Detention Act, 1908 (NSW).
Native Administration Amendment Act, 1941 (Western Australia).

A.3 Government Reports and Papers

NSW Registrar-General, Report on Vaccination, NSW Legislative Assembly, *Papers*, 1856.
Government Medical Adviser to the Colonial Secretary, 10 March 1859, NSW Legislative Assembly, *Papers*, 1858–59.
Select Committee on the Vaccination Bill, *Journal of the NSW Legislative Council*, 1872.
Select Committee: Opinions on Compulsory Vaccination, NSW Legislative Assembly, *Votes & Proceedings*, vol. 4, 1881.
Animal Vaccination: Being Information Supplied by the Government of Bombay to that of New South Wales on the Subject of Animal Lymph and Vaccination, Thomas Richards, Sydney, 1882.
Report of the Royal Commission into Management of the Quarantine Station, NSW Legislative Assembly, *Votes and Proceedings*, 1882.
Royal Commission on the Late Visitation of Small-Pox, NSW Legislative Assembly, *Votes & Proceedings*, vol. 2, 1883.
Report of the Health Officer on the Quarantine Station, North Head, NSW Legislative Assembly, *Votes & Proceedings*, vol. 2, 1883.
NSW Board of Health, Report on the Late Epidemic of Smallpox, NSW Legislative Assembly, *Votes & Proceedings*, vol. 2, 1883.
Smallpox: claims arising out of late visitation, NSW Legislative Assembly, *Votes & Proceedings*, vol. 2, 1883.
Vaccination in Darlinghurst Gaol, NSW Legislative Assembly, *Votes & Proceedings*, vol. 4, 1884.

Australasian Sanitary Conference, *Report and Minutes of Proceedings*, Government Printer, Sydney, 1884.
Norris, W. Perrin, *Report on Quarantine in Other Countries and on the Quarantine Requirements of Australia*, Government Printer, Melbourne, 1912.
Department of Public Health, Victoria, *Greenvale Sanatorium for Consumptives: Notes on Pulmonary Tuberculosis (Consumption) and on the Sanatorium Treatment of the Disease*, Government Printer, Melbourne, 1912.
Select Committee on the Prevalence of Venereal Diseases, NSW Legislative Assembly, *Votes & Proceedings*, 1915.
The Commonwealth of Australia, Department of Trade and Customs, Committee Concerning Causes of Death and Invalidity in the Commonwealth, *Report on Tuberculosis*, Government Printer, 1916.
Tropical Australia: Report of the Discussion at the Australasian Medical Congress, 1920, Government Printer, Melbourne, 1921.
Aboriginal Welfare: Initial Conference of Commonwealth and State Aboriginal Authorities, Government Printer, Canberra, 1937.

A.4 Journals and periodicals

Australasian Medical Gazette.
The British Medical Journal.
The Illustrated Sydney News.
Health: a journal dealing with developments in the field of public health in Australia.
Health and Empire.
The Empire Review.
Medical Journal of Australia.
The Lancet.
The Nineteenth Century.

A.5 Published sources – pre 1950

Anon., *Letters from a Sanatorium*, George Robertson & Co., Melbourne, 1907.
Agar, Professor W.E., *Eugenics and the Future of the Australian Population*, Brown, Prior, Anderson, Melbourne, 1939.
Beaney, J., *Vaccination and its Dangers*, R.N. Henningham, Melbourne, 1870.
Bird, S.D., *On Australasian Climates and their Influence in the Prevention and Arrest of Pulmonary Consumption*, Longman, London, 1863.
Borthwick, Thomas, *Quarantine*, Vardon & Pritchard, Adelaide, 1901.
Breinl, A. and Young, W.J., 'Tropical Australia and its Settlement', in Australian Institute of Tropical Medicine, *Collected Papers*, 3 (1922): 1–24.
Bruce, Charles, 'Mr Chamberlain and the Health of the Empire', *The Empire Review* 8 (1905): 108–21.
Buist, J.B., *Vaccinia and Variola: A Study of their Life History*, J. & A. Churchill, London, 1887.
Burnett, J. Compton, *Vaccinosis and its Cure by Thuja: With Remarks on Homoeoprophylaxis*, Homoeoepthic Publishing, London, 1897.
Campbell, Persia Crawford, *Chinese Coolie Emigration to Countries within the British Empire*, P.S. King, London, 1923.
Chidell, Fleetwood, *Australia – White or Yellow?*, Heinemann, London, 1926.

Christie, Thomas, *An Account of the Ravages Committed in Ceylon by Small-Pox, previously to the introduction of Vaccination*, J. & S. Griffith, London, 1811.
Cilento, R.W., 'Australia's Problems in the Tropics', *Report of the 21st Meeting of the Australian and New Zealand Association for the Advancement of Science*, Sydney, 1932.
——, *The White Man in the Tropics*, Government Printer, Melbourne, 1925.
Collie, Alexander, *On Fevers, Their History Aetiology, Diagnosis, Prognosis and Treatment*, H.K. Lewis, London, 1887.
Collins, William J., *Have you Been Vaccinated, and What Protection is it Against the Small Pox?*, H.K. Lewis, London, 1868.
Cook, C.E., *The Epidemiology of Leprosy in Australia*, Government Printer, Canberra, 1927.
Creighton, C., *The Natural History of Cow-Pox and Vaccinal Syphilis*, Cassell, London, 1887.
Crookshank, E.M., *The History and Pathology of Vaccination*, 2 vols, Lewis, London, 1889.
Cummins, S. Lyle, *Empire and Colonial Tuberculosis*, National Association for the Prevention of Tuberculosis, 1946.
Cumpston, J.H.L., *Quarantine: Australian Maritime Quarantine and the Evolution of International Agreements Concerning Quarantine*, Government Printer, Melbourne, 1913.
——, *The History of Small-Pox in Australia, 1788–1908*, Government Printer, Melbourne, 1914.
——, *The Health of the People: A Study in Federalism*, Roebuck, Canberra, 1978.
——, *Health and Disease in Australia: A History*, Milton Lewis (ed.), Australian Government Publishing Service, Canberra, 1989.
Cumpston, J.H.L., and F. McCallum, *The History of Small-Pox in Australia 1909–1923*, Government Printer, Melbourne, 1925.
Forward, Charles W., *The Golden Calf: An Exposure of Vaccine-Therapy*, Watkins, London, 1933.
Gregory, J.W., *Human Migration and the Future: A Study of the Causes, Effects and Control of Emigration*, Seeley, Service & Co., London, 1928.
Hayes, A.W., *Future Generations: Woman the Future Ruler of this Earth*, privately printed, Sydney, 1915.
Hutt, C.W., *International Hygiene*, Methuen, London, 1927.
Knaggs, H. Valentine, *The Truth About Vaccination: The Nature and Origin of Vaccine Lymph and the Teachings of the New Bacteriology*, Daniel, London, 1914.
McCallum, F.F., 'The Time Factor in Quarantine Practice', *Health*, 5 (1927): 45–46.
——, *International Hygiene*, Australasian Medical Publishing Co., Sydney, 1935.
Mackellar, Charles, 'Federal Quarantine', *Journal of the Royal Society of New South Wales*, 17 (1883): 278–90.
Manson, Patrick, *Tropical Diseases: A Manual of Diseases of Warm Climates*, Cassell, London, 1903.
Marchant, James, *Birth-Rate and Empire*, Williams and Norgate, 1917.
Masters, David, *The Conquest of Disease*, John Lane, London, 1925.
Morton, J., *Vaccination and its Evil Consequences*, Fuller, Parramatta, 1875.
Murray, James P., *Small-pox, Chicken-pox and Vaccination*, George Robertson, Melbourne, 1869.
Peripeteticus, A., *Cancer: A Result of Vaccination*, Stephens, Melbourne, 1898.

Price, A. Grenfell, 'The White Man in the Tropics', *Medical Journal of Australia*, 26 January 1935: 106–10.
Robinson, E., *Can Disease Protect Health? Being a Reply to Ernest Hart's pamphlet entitled 'The Truth About Vaccination'*, London, 1880.
Rogers, Sir Leonard, 'Recent Progress in the Treatment of Leprosy and its Bearing on Prophylaxis', *Proceedings of the Pan-Pacific Science Congress*, 2 (1923): 1410–18.
——, 'When Will Australia Adopt Modern Prophylactic Measures Against Leprosy?', *Medical Journal of Australia*, 18 (1930): 525–7.
Royal College of Physicians, *Report on Leprosy*, George Eyre and William Spottiswoode, London, 1867.
Ryrie, Dr G.A., *The Leper Settlement at Sungei Buloh in the Federated Malay States*, Malaya Publishing House, Singapore, 1933.
Smith, W. Ramsay, *On Consumption*, Mason, Firth & McCutcheon, Melbourne, 1909.
Thompson, J. Ashburton, 'Quarantine and Small-Pox', *Journal of the Royal Society of NSW*, 21 (1887): 227–32.
——, 'Is Leprosy a Telluric Disease', *Australasian Association for the Advancement of Science*, 6 (1895): 777–86.
——, *A Contribution to the History of Leprosy in Australia*, The New Sydenham Society, London, 1897.
——, *On the Guidance of Public Effort Towards the Further Prevention of Consumption*, Stillwell & Co., Melbourne, 1899.
Thomson, H. Hyslop, *Tuberculosis and Public Health*, Longman, Green & Co., London, 1920.
Trivett, John B., *Tuberculosis in New South Wales*, William Applegate Gullick, Sydney, 1909.
Turner, Duncan, *Is Consumption Contagious?*, Melville, Mullins & Slade, Melbourne, 1894.
Weihen, A. Wallace, 'The Medical Inspection of Immigrants to Australia', *Transactions of the Australasian Medical Congress*, 1 (1911): 637–9.
Wilkinson, H.L., *The World's Population Problems and a White Australia*, P.S. King, London, 1930.
Wilkinson, W. Camac, *Treatment of Consumption*, Macmillan, London, 1908.
Woodruff, Charles Edward, *Expansion of Races*, Rebman, London, 1909.
Woods, W. Cleaver, 'The Unsatisfactory Position of Vaccination in the Commonwealth', *Australasian Medical Gazette*, 20 May 1905, pp. 206–9.
Wright, H.P., *Leprosy: An Imperial Danger*, J. & A. Churchill, London, 1889.
Young, W.J., A. Breinl, J.J. Harris and W.Z. Osborne, 'Effect of Exercise and Humid Heat upon Pulse Rate, Blood Pressure, Body Temperature, and Blood Concentration', in Australian Institute of Tropical Medicine, *Collected Papers*, 3 (1922): 111–25.
Young, W.J., 'The Metabolism of White Races Living in the Tropics', *Annals of Tropical Medicine and Parasitology*, 9 (1915): 91–108.

B. Secondary sources

B.1 Published – post 1940

Adams, Annemarie, *Architecture in the Family Way: Doctors, Houses, and Women, 1870–1900*, McGill-Queen's University Press, 1996.

Anderson, Warwick, 'Immunities of Empire; Race, Disease and the New Tropical Medicine, 1900–1920', *Bulletin of the History of Medicine*, 70 (1996): 94–118.
——, 'Leprosy and Citizenship', *Positions*, 6 (1998): 707–30.
——, *The Cultivation of Whiteness: Science, Health and Racial Destiny in Australia*, Melbourne University Press, 2002.
Andracki, Stainslaw, *Immigration of Orientals into Canada, with Special Reference to the Chinese*, Arno Press, 1978.
Armstrong, David, 'Public Health Spaces and the Fabrication of Identity', *Sociology*, 27 (1993): 393–403.
Arnold, David, 'Smallpox and Colonial Medicine in Nineteenth Century India', in David Arnold (ed.), *Imperial Medicine and Indigenous Societies*, Manchester University Press, 1988, pp. 45–64.
——, *Colonising the Body: State Medicine and Epidemic Disease in Nineteenth Century India*, University of California Press, 1993.
——, (ed.), *Warm Climates and Western Medicine*, Rodopi, 1996.
Barta, Tony, 'Discourses of Genocide in Germany and Australia: A Linked History', *Aboriginal History*, 25 (2001): 37–56.
Bashford, Alison, 'Female Bodies at Work: Gender and the Re-forming of Colonial Hospitals', *Australian Cultural History*, 13 (1994): 65–81.
——, 'Separatist Health: Meanings of Women's Hospitals in England and Australia, c. 1870–1930', in Lilian R. Furst (ed.), *Climbing a Long Hill: Women Healers and Physicians*, University Press of Kentucky, 1997, pp. 198–220.
——, *Purity and Pollution: Gender, Embodiment and Victorian Medicine*, Macmillan, 1998.
——, ' "Is White Australia Possible?" Race, Colonialism and Tropical Medicine', *Ethnic and Racial Studies*, 23 (2000): 248–71.
——, 'Tuberculosis and Economy: Public Health and Labour in the Early Welfare State', *Health and History*, 4 (2002): 19–40.
Bashford, Alison and Carolyn Strange, 'Asylum Seekers and National Histories of Detention', *Australian Journal of Politics and History*, 48 (2002): 509–27.
Bashford, Alison and Carolyn Strange, 'Isolation and Exclusion in the Modern World', in Carolyn Strange and Alison Bashford (eds), *Isolation: Places and Practices of Exclusion*, Routledge, 2003 pp. 1–19.
Bell, Heather, *Frontiers of Medicine in the Anglo-Egyptian Sudan 1899–1940*, Clarendon Press, 1999.
Bland, Lucy and Laura Doan, (eds), *Sexology in Culture: labelling bodies and desires*, University of Chicago Press, 1998.
Bock, Gisela and Pat Thane (eds), *Maternity and Gender Policies: Women and the Rise of the European Welfare States, 1880s–1950s*, Routledge, 1991.
Bowers, J.Z., 'The Odyssey of Smallpox Vaccination', *Bulletin of the History of Medicine*, 55 (1981): 17–33.
Brawley, S., *The White Peril: Foreign Relations and Asian Immigration to Australasia and North America 1919–1978*, University of New South Wales Press, 1995.
Brown, JoAnne, 'Purity and Danger in Colour: Notes on Germ Theory and the Semantics of Segregation, 1895–1915', in Jean-Paul Gaudillière and Ilana Löwy (eds), *Heredity and Infection: the History of Disease Transmission*, Routledge, 2001, pp. 101–32.
Bryder, Linda, *Below the Magic Mountain: A Social History of Tuberculosis in Twentieth-Century Britain*, Clarendon Press, 1988.

——, ' "A Health Resort for Consumptives": Tuberculosis and Immigration to New Zealand, 1880-1914', *Medical History*, 40 (1996): 453-71.
Buckingham, Jane, *Leprosy in Colonial South India: Medicine and Confinement*, Palgrave Macmillan, 2002.
Bunton, Robin and Roger Burrows, 'Consumption and Health in the "Epidemiological" Clinic of Late Modern Medicine', in Robin Bunton, Sarah Nettleton and Roger Burrows (eds), *The Sociology of Health Promotion: Critical Analyses of Consumption, Lifestyle & Risk*, Routledge, 1995, pp. 206-22.
Burnard, Trevor, ' "The Countrie Continues Sicklie": White Mortality in Jamaica, 1655-1780', *Social History of Medicine*, 12 (1999): 45-72.
Burnet, Sir Macfarlane, 'Biology and Medicine', *The Eugenics Review*, 49 (1957): 127-35.
——, 'Migration and Race Mixture from the Genetic Angle', *The Eugenics Review*, 51 (1959): 93-7.
Bynam, W.F., 'Policing Hearts of Darkness: Aspects of the International Sanitary Conferences', *History and Philosophy of the Life Sciences*, 15 (1993): 421-34.
Campbell, Judy, *Invisible Invaders: Smallpox and other Diseases in Aboriginal Australia 1780-1880*, Melbourne University Press, 2002.
Castel, Robert, 'From Dangerousness to Risk', in Graham Burchell *et al.* (eds), *The Foucault Effect: Studies in Governmentality*, University of Chicago Press, 1991, pp. 281-98.
Chesterman, J. and B. Galligan, *Citizens Without Rights: Aborigines and Australian Citizenship*, Cambridge University Press, 1997.
Cohn, B.S., *Colonialism and its Forms of Knowledge: The British in India*, Princeton University Press, 1996.
Coker, Richard J., *From Chaos to Coercion: Detention and the Control of Tuberculosis*, St Martin's Press, 2000.
——, 'Civil Liberties and Public Good: Detention of Tuberculous Patients and the Public Health Act 1984', *Medical History*, 45 (2001): 339-56.
Coons, Ronald E., 'Steamships and Quarantines at Trieste, 1837-1848', *Journal of the History of Medicine and Allied Sciences*, 44 (1989): 28-55.
Corris, Peter, ' "White Australia" in Action: The Repatriation of Pacific Islanders from Queensland', *Historical Studies*, 15 (1972): 237-50.
Craddock, Susan, 'Sewers and Scapegoats: Spatial Metaphors of Smallpox in Nineteenth Century San Francisco', *Social Science and Medicine*, 7 (1995): 957-68.
Craddock, Susan and Michael Dorn, 'Nationbuilding: Gender, Race and Medical Discourse', *Journal of Historical Geography*, 27 (2001): 313-18.
Crotty, Martin, John Germov and Grant Rodwell (eds), *'A Race for a Place': Eugenics, Darwinism and Social Thought and Practice in Australia*, Faculty of Arts and Social Sciences, University of Newcastle, 2000.
Curson, P.H., *Times of Crisis: Epidemics in Sydney 1788-1900*, Sydney University Press, 1985.
Curson, Peter and Kevin McCracken, *Plague in Sydney: The Anatomy of an Epidemic*, University of New South Wales Press, 1989.
Curthoys, Ann, 'Eugenics, Feminism and Birth Control: The Case of Maion Piddington', *Hecate*, 15 (1989): 73-89.
——, 'Expulsion, Exodus and Exile in White Australian Historical Mythology', in Richard Nile and Michael Williams (eds), *Imaginary Homelands: The Dubious*

Cartographies of Australian Identity, University of Queensland Press, 1999, pp. 1–18.
Curtin, Philip D., *Death by Migration: Europe's Encounter with the Tropical World in the Nineteenth Century*, Cambridge University Press, 1989.
Harriet Deacon, 'Leprosy and Racism at Robben Island', *Studies in the History of Cape Town*, 7 (1994): 45–83.
——, 'Racial Segregation and Medical Discourse in Nineteenth Century Cape Town', *Journal of Southern African Studies*, 22 (1996): 187–308.
——, 'Racism and Medical Science in South Africa's Cape Colony in the mid- to late Nineteenth Century', *Osiris*, 15 (2000): 190–206.
Dean, Mitchell, *The Constitution of Poverty: Toward a Genealogy of Liberal Governance*, Routledge, 1991.
——, *Governmentality: Power & Rule in Modern Society*, Sage, 1999.
Douglas, Mary, *Purity and Danger: An Analysis of the Concepts of Pollution and Taboo*, Routledge, 1994.
Durbach, Nadja, ' "They Might as Well Brand Us": Working-Class Resistance to Compulsory Vaccination in Victorian England', *Social History of Medicine*, 13 (2000): 45–61.
Dutton, David, *One of Us? A Century of Australian Citizenship* University of New South Wales Press, 2002.
Evans, Raymond, Kay Saunders and Kathryn Cronin, *Exclusion, Exploitation and Extermination: Race Relations in Colonial Queensland*, Australia and New Zealand Book Co., 1975.
Eyler, John, 'Scarlet Fever and Confinement: The Edwardian Debate over Isolation Hospitals,' *Bulletin of the History of Medicine*, 61 (1987): 1–24.
——, *Sir Arthur Newsholme and State Medicine 1885–1935*, Cambridge University Press, 1997.
Fenn, Elizabeth A., *Pox Americana: The Great Smallpox Epidemic of 1775–82*, Hill & Wang, 2001.
Finnane, Mark (ed.), *Policing in Australia: Historical Perspectives*, University of New South Wales Press, 1987.
Foley, Jean Duncan, *In Quarantine: A History of Sydney's Quarantine Station 1828–1984*, Kangaroo Press, 1995.
Foucault, Michel, 'About the Concept of the Dangerous Individual in 19th Century Legal Psychiatry', in David. N. Weisstub (ed.), *Law and Psychiatry*, Pergamon Press, 1978.
——, 'The Politics of Health in the Eighteenth Century', in Paul Rabinow (ed.), *The Foucault Reader*, Pantheon Books, 1984, pp. 273–89.
——, 'Governmentality', in Graham Burchell *et al.* (eds), *The Foucault Effect*, University of Chicago Press, 1991.
——, *Discipline and Punish*, Penguin, 1991.
——, 'The Birth of Social Medicine', in James D. Faubion (ed.), *Essential Works of Michel Foucault*, Vol. 3, 'Power', New Press, 2000, pp. 137–42.
Garton, Stephen, 'Policing the Dangerous Lunatic: Lunacy Incarceration in New South Wales, 1843–1914', in Mark Finnane (ed.), *Policing in Australia: Historical Perspectives*, University of New South Wales Press, 1987, pp. 74–87.
——, *Medicine and Madness: A Social History of Insanity in New South Wales 1880–1940*, University of New South Wales Press, 1988.
——, *Out of Luck: Poor Australians and Social Welfare*, Allen & Unwin, 1990.

——, 'Sound Minds and Healthy Bodies: Re-considering Eugenics in Australia, 1914–1940', *Australian Historical Studies*, 26 (1994): 163–81.
Gaudillière, Jean-Paul and Ilana Löwy (eds), *Heredity and Infection: The History of Disease Transmission*, Routledge, 2001.
Gillespie, James A., *The Price of Health: Australian Governments and Medical Politics 1910–1960*, Cambridge University Press, 1991.
Hacking, Ian, 'How Should We Do a History of Statistics', in Burchell *et al.* (eds), *The Foucault Effect: Studies in Governmentality*, University of Chicago Press, 1991, pp. 181–96.
Hall, Lesley, 'Feminist reconfigurations of heterosexuality in the 1920s', in Lucy Bland and Laura Doan (eds), *Sexology in culture: labelling bodies and desires*, University of Chicago Press, 1998, pp. 135–49.
Hamlin, Christopher, 'State Medicine in Great Britain', in Dorothy Porter (ed.), *The History of Public Health and the Modern State*, Rodopi, 1994, pp. 132–64.
Hardy, Anne, *The Epidemic Streets: Infectious Disease and the Rise of Preventive Medicine, 1856–1900*, Clarendon Press, 1993.
Harrison, Mark, *Public Health in British India: Anglo-Indian Preventive Medicine, 1859–1914*, Cambridge University Press, 1994.
——, *Climates and Constitutions: Health, Race, Environment and British Imperialism in India 1600–1850*, Oxford University Press, 1999.
Haynes, Douglas M., *Imperial Medicine: Patrick Manson and the Conquest of Tropical Disease*, University of Pennsylvania Press, 2001.
Hooker, Claire and Alison Bashford, 'Diphtheria and Australian Public Health: Bacteriology and its Complex Applications, c.1890–1930', *Medical History*, 46 (2002): 41–64.
Hooker, Claire, 'Sanitary Failure and Risk: Pasteurisation, Immunisation and the Logics of Prevention', in Alison Bashford and Claire Hooker (eds), *Contagion: Historical and Cultural Studies*, Routledge, 2001, pp. 129–49.
Howard-Jones, N., *The Scientific Background of the International Sanitary Conferences, 1851–1938*, World Health Organization, 1975.
Huttenback, R.A., 'No Strangers Within the Gates: Attitudes and Policies towards the non-white residents of the British Empire of Settlement', *Journal of Imperial and Commonwealth History*, 1 (1972–3): 271–302.
Hyslop, Anthea, 'Old Ways, New Means: Fighting Spanish Influenza in Australia, 1918–1919', in Linda Bryder and Derek A. Dow (eds), *New Countries, Old Medicine*, Pyramid Press, 1995, pp. 46–53.
——, 'Insidious Immigrant: Spanish Influenza and Border Quarantine in Australia, 1919', in S. Barry and B. Reid (eds), *Migration to Mining*, Northern Territory University Press, 1997.
——, 'A Question of Identity: J.H.L Cumpston and Spanish Influenza, 1918–1919', *Australian Cultural History*, 16 (1997/98): 60–78.
Irving, Helen, *To Constitute a Nation: A Cultural History of Australia's Constitution*, Cambridge University Press, 1999.
Jebb, Mary Ann, 'The Lock Hospitals Experiment: Europeans, Aborigines and Venereal Disease', *European-Aboriginal Relations in Western Australian History*, 8 (1984): 68–87.
Jones, Ross L., 'The Master Potter and the Rejected Pots: Eugenic Legislation in Victoria 1918–1939', *Historical Studies*, 113 (1999): 319–42.

Kalpagam, U., 'The Colonial State and Statistical Knowledge', *History of the Human Sciences*, 13 (2000): 37–55.
Kern, Stephen, *The Culture of Time and Space, 1880–1918*, Harvard University Press, 1983.
Kociumbas, Jan, 'Reflecting on "the Century of the Child": Child Study and the School Medical Service in New South Wales', in Crotty *et al.* (eds), *'A Race for a Place'*, pp. 221–8.
Leavitt, Judith Walzer, *Typhoid Mary: Captive to the Public's Health*, Beacon Press, 1996.
Levine, Philippa, *Prostitution, Race and Politics: Policing Venereal Disease in the British Empire*, Routledge, 2003.
Lewis, Milton, 'Introduction', in J.H.L. Cumpston, *Health and Disease in Australia: A History*, Australian Government Printing Service, 1989.
——, *Thorns on the Rose: The History of Sexually Transmitted Diseases in Australia in International Perspective*, Australian Government Publishing Service, 1998.
Livingstone, David N., 'Human Acclimatization: Perspectives on a Contested Field of Inquiry in Science, Medicine and Geography', *History of Science*, 25 (1987): 359–94.
——, 'Tropical Climate and Moral Hygiene: The Anatomy of a Victorian Debate', *British Journal of the History of Science*, 32 (1999): pp. 93–110.
Ludmerer, Kenneth M, 'Genetics, Eugenics and the Immigration Restriction Act of 1924' *Bulletin of the History of Medicine*, 46 (1972): 59–81.
Lupton, Deborah, *The Imperative of Health: Public Health and the Regulated Body*, Sage, 1995.
——, *Risk*, Routledge, 1999.
Lux, Maureen K., *Medicine that Walks: Disease, Medicine and Canadian Plains Native People 1880–1940*, University of Toronto Press, 2001.
McGregor, Russell, ' "Breed Out the Colour" or the Importance of Being White', *Australian Historical Studies*, 120 (2002): 297–301.
Macleod, R.M., 'Law, Medicine and Public Opinion: The Resistance to Compulsory Health Legislation, 1870–1907', *Public Law*, Parts I and II (1967): 107–28, 189–211.
Macleod, Roy and Milton Lewis (eds), *Disease, Medicine and Empire: Perspectives on Western Medicine and the Experience of European Expansion*, Routledge, 1988.
McSherry, Bernadette, ' "Dangerousness" and Public Health', *Alternative Law Journal*, 57 (1998): 276–80.
Manderson, Lenore, 'Wireless Wars in the Eastern Arena', in Paul Weindling (ed.), *International Health Organisations*, Cambridge University Press, 1995, pp. 109–33.
——, *Sickness and the State: Health and Illness in Colonial Malaya, 1870–1940*, Cambridge University Press, 1996.
Markel, Howard, ' "Knocking out the Cholera": Cholera, Class and Quarantines in New York City, 1892', *Bulletin of the History of Medicine*, 69 (1995): 420–57.
Markus, Andrew, *Fear and Hatred: Purifying Australia and California 1850–1901*, Hale & Iremonger, 1979.
——, *Australian Race Relations*, Allen & Unwin, 1994.
Martin, Emily, 'Toward an Anthropology of Immunology: The Body as Nation-State', *Medical Anthropology Quarterly*, 4 (1990): 410–26.
Mawani, Renisa, 'Legal Geographies of Aboriginal Segregation in British Columbia: The Making and Unmaking of the Songhees Reserve, 1850–1911',

in Carolyn Strange and Alison Bashford (eds), *Isolation: Places and Practices of Exclusion*, Routledge, 2003, pp. 173–90.

——, ' "The Island of the Unclean": Race, Colonialism and "Chinese Leprosy" in British Columbia, 1891–1924', *Journal of Law, Social Justice and Global Development* (2003), http://elj. warwick.ac.uk/global/

Mayne, Alan, *Fever, Squalor and Vice: Sanitation and Social Policy in Victorian Sydney*, University of Queensland Press, 1982.

——, 'The Dreadful Scourge': Responses to Smallpox in Sydney and Melbourne, 1881–2', in Roy Macleod and Milton Lewis (eds), *Disease, Medicine and Empire*, Routledge, 1988, pp. 219–41.

Mehta, Uday S., 'Liberal Strategies of Exclusion', in Frederick Cooper and Ann Laura Stoler (eds), *Tensions of Empire: Colonial Cultures in a Bourgeois World*, University of California Press, 1997, pp. 59–86.

Mooney, Graham, 'Public Health versus Private Practice: The Contested Development of Compulsory Infectious Disease Notification in Late-Nineteenth-Century Britain', *Bulletin of the History of Medicine*, 73 (1999): 238–67.

Moses, A. Dirk, 'Conceptual Blockages and Definitional Dilemmas in the "Racial Century": Genocides of Indigenous Peoples and the Holocaust', *Patterns of Prejudice*, 36 (2002): 7–36.

Murnane, Mary and Kay Daniels, 'Prostitutes as "Purveyors of Disease": Venereal Disease Legislation in Tasmania, 1868–1945', *Hecate*, 5 (1979): 5–21.

O'Connor, Erin, *Raw Material: Producing Pathology in Victorian Culture*, Duke University Press, 2000.

Osborne, Thomas, 'Security and Vitality: Drains, Liberalism and Power in the Nineteenth Century', in Andrew Barry *et al.* (eds), *Foucault and Political Reason: Liberalism, Neo-Liberalism and Rationalities of Government*, University of Chicago Press, 1996, pp. 99–122.

Otis, Laura, *Membranes: Metaphors of Invasion in Nineteenth-Century Literature, Science and Politics*, Johns Hopkins University Press, 1998.

Ott, Katherine, *Fevered Lives: Tuberculosis in American Culture since 1870*, Harvard University Press, 1996.

Peers, Douglas, 'Soldiers, Surgeons and the Campaigns to Combat Sexually Transmitted Diseases in Colonial India, 1805–1860', *Medical History*, 42 (1998): 137–60.

Pelling, Margaret, 'The Meaning of Contagion: reproduction, medicine and metaphor', in Alison Bashford and Claire Hooker (eds), *Contagion: Historical and Cultural Studies*, Routledge, 2001, pp. 15–38.

Petersen, Alan, 'Risk, Governance and the New Public Health', in Alan Petersen and Robin Bunton (eds), *Foucault, Health and Medicine*, Routledge, 1997, pp. 189–206.

Petersen, Alan and Deborah Lupton. *The New Public Health: Health and Self in the Age of Risk*, Allen & Unwin, 1996.

Poovey, Mary, *Making a Social Body: British Cultural Formation*, University of Chicago Press, 1995.

Porter, Dorothy, ' "Enemies of the Race": Biologism, Environmentalism and Public Health in Edwardian England,' *Victorian Studies*, 34 (1991): 159–78.

——, (ed.), *The History of Public Health and the Modern State*, Rodopi, 1994.

——, *Health, Civilization and the State: A History of Public Health from Ancient to Modern Times*, Routledge, 1999.

Porter, Dorothy and Roy Porter, 'The Politics of Prevention: Anti-Vaccinationism and Public Health in Nineteenth Century England', *Medical History*, 32 (1988): 231–52.
——, 'The Enforcement of Health: The British Debate', in Elizabeth Fee and Daniel M. Fox (eds), *AIDS: The Burdens of History*, University of California Press, 1988, pp. 97–120.
Powell, J.M., 'Medical Promotion and the Consumptive Immigrant to Australia', *Geographical Review*, 63 (1973): 449–76.
Pringle, Rosemary, 'Octavius Beale and the Ideology of the Birth-Rate', *Refractory Girl*, 3 (1973): 19–27.
Proctor, Robert, *Racial Hygiene: Medicine Under the Nazis*, Harvard University Press, 1988.
——, 'The Destruction of "Lives Not Worth Living" ', in Jennifer Terry and Jacqueline Urla (eds), *Deviant Bodies: Critical Perspectives on Difference in Science and Popular Culture*, Indiana University Press, 1995, pp. 170–96.
Proust, A.J., 'The Invalid Pension and Sickness Benefits in Australia prior to 1948', in A.J. Proust (ed.), *History of Tuberculosis in Australia, New Zealand and Papua New Guinea*, Brolga Press, 1991.
Roe, Michael, 'The Establishment of the Australian Department of Health: Its Background and Significance', *Historical Studies*, 17 (1976): 176–92.
——, *Nine Australian Progressives: Vitalism in Bourgeois Social Thought*, University of Queensland Press, 1984.
——, *Australia, Britain and Migration, 1915–1940*, Cambridge University Press, 1995.
——, *Life over Death: Tasmanians and Tuberculosis*, Tasmanian Historical Research Association, 1999.
Rose, Nikolas, *Governing the Soul: The Shaping of the Private Self*, Routledge, 1990.
——, 'Medicine, History and the Present', in Colin Jones and Roy Porter (eds), *Reassessing Foucault: Power, Medicine and the Body*, Routledge, 1994.
——, *Powers of Freedom: Reframing Political Thought*, Cambridge University Press, 1999.
Rosen, George, 'Cameralism and the Concept of Medical Police', *Bulletin of the History of Medicine*, 27 (1953): 21–42.
——, *A History of Public Health*, MD Publications, 1958.
Rosenberg, Charles, *Explaining Epidemics and other Studies in the History of Medicine*, Cambridge University Press, 1992.
Saunders, Kay and Helen Taylor, ' "To Combat the Plague": The Construction of Moral Alarm and State Intervention in Queensland During World War II', *Hecate*, 14 (1988): 5–30.
Saunders, Suzanne, 'Isolation: the Development of Leprosy Prophylaxis in Australia', *Aboriginal History*, 14 (1990): pp. 168–81.
Sears, Alan, ' "The Teach Them How to Live": The Politics of Public Health from Tuberculosis to AIDS', *Journal of Historical Sociology*, 5 (1992): 70–71.
Shah, Nayan, *Contagious Divides: Epidemics and Race in San Francisco's Chinatown*, University of California Press, 2002.
Sibley, David, *Geographies of Exclusion: Society and Difference in the West*, Routledge, 1995.
Smart, Judith, 'Sex, the State and the "Scarlet Scourge": Gender, Citizenship and Venereal Disease Regulation in Australia during the Great War', *Women's History Review*, 7 (1998): 5–36.

Smith, F.B., *The Retreat of Tuberculosis 1850–1950*, Croom Helm, 1988.

Soloway, Richard, *Demography and Degeneration: Eugenics and the Declining Birthrate in Twentieth Century Britain*, University of North Carolina Press, 1995.

Stern, Alexandra Minna, 'Buildings, Boundaries and Blood: Medicalization and Nation-Building on the US-Mexico Border, 1910–1930', *Hispanic American Historical Review*, 79 (1999): 41–81.

Stoler, Ann Laura, 'Rethinking Colonial Categories: European Communities and the Boundaries of Rule', *Comparative Studies in Society and History*, 31 (1989): 134–61.

——, *Race and the Education of Desire: Foucault's History of Sexuality and the Colonial Order of Things*, Duke University Press, 1995.

——, 'Sexual Affronts and Racial Frontiers: European Identities and the Cultural Politics of Exclusion in Colonial Southeast Asia', in Frederick Cooper and Ann Laura Stoler (eds), *Tensions of Empire: Colonial Cultures in a Bourgeois World*, University of California Press, 1997.

——, 'Making Empire Respectable: The Politics of Race and Sexual Morality in Twentieth-Century Colonial Cultures', in A. McClintock, A. Mufti and E. Shoat (eds), *Dangerous Liaisons: Gender, National and Postcolonial Perspectives*, University of Minnesota Press, 1997.

Strange, Carolyn and Tina, Loo *Making Good: Law and Moral Regulation in Canada, 1867–1939*, University of Toronto Press, 1997.

Tauber, Alfred, *The Immune Self: Theory or Metaphor*, Cambridge University Press, 1994.

Thomson, Matthew, *The Problem of Mental Deficiency: Eugenics, Democracy and Social Policy in Britain, c.1870–1959*, Oxford University Press, 1998.

Thomas, Nicholas, *Colonialism's Culture: Anthropology, Travel and Government*, Polity, 1994.

Thongchai, Winichakul, *Siam Mapped: A History of the Geo-Body of a Nation*, University of Hawaii Press, 1994.

Torpey, John, *The Invention of the Passport: Surveillance, Citizenship and the State*, Cambridge University Press, 2000.

Waldby, Catherine, *AIDS and the Body Politic: Biomedicine and Sexual Difference*, Routledge, 1996.

Walker, David, 'Climate, Civilization and Character in Australia, 1880–1940', *Australian Cultural History*, 16 (1997/98): 77–95.

——, *Anxious Nation: Australia and the Rise of Asia, 1850–1939*, University of Queensland Press, 1999.

Watters, Greg, 'The *S.S. Ocean*: Dealing with Boat People in the 1880s', *Australian Historical Studies*, 120 (2002): 331–43.

Weindling, Paul, 'Public Health in Germany', in Dorothy Porter (ed.), *The History of Public Health and the Modern State*, Rodopi, 1994, pp. 119–31.

——, (ed.), *International Health Organisations and Movements, 1918–1939*, Cambridge University Press, 1995.

Williams, Naomi, 'The Implementation of Compulsory Health Legislation: Infant Smallpox Vaccination in England and Wales, 1840–1890', *Journal of Historical Geography*, 20 (1994): 396–412.

Worboys, Michael, 'Manson, Ross and Colonial Medical Policy: Tropical Medicine in London and Liverpool, 1899–1914', in Roy Macleod and

Milton Lewis (eds), *Disease, Medicine and Empire*, Routledge, 1998, pp. 21–37.

——, 'The Colonial World as Mission and Mandate: Leprosy and Empire, 1900–1940', *Osiris*, 15 (2000): 207–20.

——, *Spreading Germs: Disease Theories and Medical Practice in Britain, 1865–1900*, Cambridge University Press, 2000.

Yarwood, A.T., *Asian Migration to Australia: The Background to Exclusion, 1896–1923*, Melbourne University Press, 1964.

——, 'The Overseas Indians: A Problem in Indian and Imperial Politics at the end of World War One', *Australian Journal of Politics and History*, 14 (1968): 204–18.

——, 'Sir Raphael Cilento and the *White Man in the Tropics*', in Roy Macleod and Donald Denoon (eds), *Health and Healing in Tropical Australia*, James Cook University Press, 1991, pp. 47–63.

Yarwood, A.T., and M.J. Knowling, *Race Relations in Australia: A History*, Methuen Australia, 1982.

B.2 Theses

Allan, R. Tennyson, 'Leprosy at Nauru, Central Pacific', Doctor of Medicine thesis, University of Melbourne, 1939.

Elder, Catriona, 'Dreams and Nightmares of a White Australia', PhD thesis, Australian National University, 1999.

Hardy, P. Susan, ' "Surgical Spirit": Listerism in New South Wales', PhD thesis, University of New South Wales, 1990.

Deacon, Harriet, 'A history of the medical institutions on Robben Island, 1846–1910', DPhil thesis, University of Cambridge, 1994.

Nugent, Maria, 'Revisiting La Perouse: a postcolonial history', University of Technology, Sydney, 2001.

Thame, Claudia, 'Health and the State in Australia', PhD thesis, Australian National University, 1974.

Ussher, Greg, 'The "medical gaze" and the "watchful eye": the prevention, treatment and epidemiology of venereal diseases in NSW, 1900–1925', PhD thesis, University of Sydney, forthcoming.

Index

Aboriginal people: coercive treatment of 98, 104, 172; deaths from smallpox 41; detention in sanatoria 69; idea of impending extinction of 106, 109, 149; leprosy and treatment of 12, 94, 95, 96, 98, 99, 100, 103, 104, 105, 106, 108–9, 113; mortality from tuberculosis 59; protection legislation 169; reserves 104–5, 106, 168–9; theories of sociality 103–4; travel restrictions 107, 113
abortion 173
An Account of the Ravages Committed in Ceylon by Small-Pox 31
administration: colonial 1, 9, 115, 120; Commonwealth of Australia 121–4; and making boundaries 6; national borders and quarantine 123–4, 135–6
aetiology 15, 63, 65, 84, 158
Africa 23, 91
African-Americans 64
Agar, W.E. 145
AIDS *see* HIV/AIDS
air travel 133
Albany 125, 126
alcoholics and alcoholism 77, 174, 184
Alice Springs 105
alien-ness 1, 162
Aliens Order (Britain) 146
Anatomy Act (1831) 33, 52
Anderson, Warwick 91, 161
animal disease: and smallpox 19–20, 23
Animal Vaccination (Richards) **26–7**
anorexics 71
anthropological studies 5–6, 6–7
anti-vaccinationists 17–18, 19, 20, 22, 52, 69, 128–9
antibiotics 2, 105
Antigua 83, 85

architecture 43, 71, 87, 92
Armstrong, David 10–11, 48
Arthur, Richard 169, 174, 183
asepsis 46–7
Asia 126, 129, 131, 136
Asian people 4, 60, 82, 153–4, 154, 161, *see also* Chinese people
assimilation: programmes of 109, 110, 113, 141, 172
asylum seekers: detention of 170
asylums 60, 66, 76, 86, 87, 91, 155, 167, 181
Australasian colonies 3–4, 12, 25, 29, 59, 120, 130, 147
Australasian Medical Congress 105–6, 139, 159
Australia: creation of Commonwealth 116, 120; decline in women's reproductivity 172–3; eugenics and public health 164, 165, 182–4; imagined as 'virgin' and uncontaminated 13, 110, 126, **127**, 131; as island-nation 13, 116, 124–8, 129, 130, 136; quarantine as central to 116, 121, 124–5, 126, 128, 135–6, 162, 188; racial mapping of cleanliness and contamination 49, 126, 148; venereal disease policy 168–70, 171–2; white people in tropics 6, 110–13, 140, 144, 157, 157–63; as white settler society and colonising nation 1, 3–4, 9–10, 89, 141, 148, 186–7
Australia for the Consumptive Invalid 63
Australia – white or yellow? (Chidell) 157, 161
Australian colonies 3–4, 119, 120, 128; compulsory vaccination 51–3; English emigrants to 36–8; status of smallpox 22, 40
Australian Institute of Tropical Medicine 122, 159, 160

246

'auto-inoculation'   75, 77

bacteriology: ideas about leprosy   84, 96; importance of tuberculosis 64–5; refinement of diagnoses   61, 62
Bahamas   87
Balibar, Etienne   5, 111
Barbados   85
Barrett, J.W.   160
barriers *see* lines or barriers
Beaney, J.W.   20
Beer, Gillian   133
Bell, Heather   124
Bentham, Jeremy   8, 92
Berlin International Congress on Hygiene and Demography   122
Bermuda   85
bills of health   141–2, 174
biomedical discourse   4, 129, 165
biopolitics   8, 42–3, 119, 181; colonial and international   9–10, 113, 137–8, 142; management of migration   146–7; quarantine and immigration   139, 162; reproduction of the population   172, 173, 180; sex, health and population   165, 166; vaccination 33, 34
birth control   164, 172, 173, 174; voluntary sterilisation   182
birth rates: concerns   164–5, 166, 172
black people: Canadian exclusion of 144
Blackwell, Elizabeth   19
blood: and immunity   18, 74; and sexuality   166
bodies   4; and boundaries   12, 115; circulation of disease through   44; governance of in management of diseases   12; in quarantined space 45; training of consumptives in sanatoria   62, 71–2, 72–4; vaccination   18, 23, 34
body: social   4, 34, *see also* 'geo-body' of Australia
Bombay   25, **27**, 83
borders   1, 5, 6; between infected and uninfected   40, 161; medico-legal control of   13, 115, 124, 142, 151, 152; membrane-line of skin 15, 38, 124; national   6, 12, 38, 110, 115, 123–9, 135–6, 137; public health   12; quarantine   36, 38, 40, 115, 123–9, 135–6, 182; screening of Britons   139, 152–3, 155; segregation   38, *see also* lines or barriers
boundaries   1, 113; crossed by vaccination   16, 23, 38; geographic, legal and actual   6, 48, 115; imperially and globally regulated   12; interior *see* interior frontiers; 'leper line'   106–7, 161; national   123, 124; quarantine as policing of   123, 135; racial   6, 110; of rule   1, 5–6, 13, 93, 114, 162
Breinl, Anton   159, 160
Brereton, John le Gay   32
Brisbane   159, 168
Britain   8, 136, 157, 177, 178; anti-CD Act agitation   167, 170; anxieties about vaccination   20, 21–2; cholera   41; eugenics   141, 146; interwar social hygiene   164; medical and public health developments   2–3, 3, 10; notification of infectious diseases   61; treatment of consumptives   63, 64; trend away from public health detention   170–1
British colonialism   1, 81, 117; and leper enclosure   90–1
British colonies: and creation of Commonwealth of Australia   116, 120, 125; development of social policy   8; and international health networks   142; management of lepers   86; settlement   138, 141, 155; transportation to   3, *see also* Australasian colonies; Straits Settlements
British Columbia   81, 88, 89, *see also* D'Arcy Island
British dominions   2, 3, 146, 157, 170
British Empire Leprosy Relief Association   89, 90

British Eugenics Education Society 146, 174, 174–7
British Guiana 83
British people: and 'alien' diseases 14, 59, 96; border screening of 139–40, 152–3, 153–4, 155; 'coloured' subjects 149; in imagined geography of Australia 6, 147; migration 81, 104, 146–7; and tuberculosis 59
British Social Hygiene Council 154, 170–1, 172, 173–4, 180
Brown, Isaac 63
Bryder, Linda 63, 70
Burchell, Graham 8
bureaucracy: health and tropical hygiene 158–9; immigration 151, 153; of public health 2, 12, 36, 39, 40–1, 42, 43–5, 116, 173; quarantine 134; significance of borders 123–4, 135; vaccination 33, 34
Bynam, W.F. 124–5

Caffyn, S. Mannington 41, 50
Calcutta 83, 85
Campbell, Persia Crawford 149
Canada 3, 24, 36, 38, 88, 89, 143, 144, 146, 177, *see also* British Columbia; New Brunswick
cancer 21
Canterbury, Dean of 156, 161
Cape Colony 83, 85, 87, 88, 107, *see also* Robben Island
carceral spaces 1, 12, 39, 68, 88, 167, 169, 170
Castel, Robert 60
certification: health 153, 174, 177; labourers 151; vaccination 34, 36
Ceylon 28–9, 120, 131; leprosy 83, 87, 88; smallpox and vaccination 24, 30–2, **31**
Chadwick, Edwin 8, 130
Chamberlain, Joseph 83, 149–50
Chidell, Fleetwood 157, 161
children: in sanatoria 73–4; and vaccination 4, 5, 16, 17, 20, 25, 29–32, 50, 52

China 23, 88, 111–12, 131
Chinese people: association of with leprosy 88–9, 95, 96, 99, 108, 110, 142–3, 148; forced onto leper colonies 12, 94; immigrants to Australasian colonies 147–8; laws of exclusion and restriction 88, 141, 143, 148, 149, 151–2; leprosy cases in New South Wales 96–7; movement of diaspora 22, 59, 81, 82, 88–9; as 'others' in Australia 4, 110, 113, 138, 139, 157; in Quarantine Station 49, 50, 54, 56–7; view of leprosy 86
cholera 41, 131, 135, 153; emergency quarantine measures 39, 117, 118, 119, 167; European epidemics 15, 117, 119, 142; maps of distribution 126, 130; prevention measures in modern period 133–4
Christie, Thomas 31
Cilento, Raphael 94, 100, 102, 104, 105, 110, 123, 158, 159
circulations 2, 115, 185; of contagious matter 15, 16, 113; epidemic 44–5; of goods 81, 84
citizenship 3, 12; cultivation of in leper colony 91; cultivation of in sanatoria 12, 13, 62, 70–9, 80; exclusion of Aboriginal people from 103–4; hygienic 62, 77–9, 103; identities 1, 102, 147; and making boundaries 6; and public health 11, 21, 77, 113, 116–17, 170–80, 189
civic responsibility 62, 77–8, 80, 91, 93, 99, 102, 104, 173, 177, 180, 185
class: mapping of cleanliness and contamination 49; vaccination and crossing of boundaries 29, *see also* working classes
classification: eugenics and segregation 184, 187; international hygiene 142–3, 147; social 6, 48–9, 104, 166–7
cleanliness: and contamination 46–7, **47**, 48–9, 84; Cumpston's ideas 187–9; and imagining of

Australia   126, 139, 147, 150; imperial   1, 5, 84, 113, 188
climate: and race   140, 144; and tropical medicine   157-8, 160
coercion   7, 92, 186; of Aboriginal people   98, 104, 172; confinement in sanatoria   62-3, 69, 79; practices and places of   11, 82, 171; for smallpox vaccination   53-4, 59
Colombo   131
Colonial Office   142
colonialism: in Australian history   3-4, 9-10; boundaries of rule   5-6, 13, 114; and contagion   14, 15; importance of statistics   33; leprosy policy and management   81-2, 86-7, 90-1, 93, 98; management of diseases   11-12, 24; mapping of cleanliness and contamination   49, 148; and medicine   2; obsessive pursuit of whiteness   140-1, 163; roles of public health and hygiene   7, 9-10, 187-9; segregation and race management   107; and tropical medicine   157-8; and vaccination   15, 16, 25, 34, 38, *see also* British colonialism
colonisation   8, 12; and Australian history   3-4, 9-10, 112, 163; and contagion   15, 106; and imperial hygiene   113, 188, 189; and vaccination   25, 38
colour bar   182
'coloured people': and Australian racial politics   148, 162; British policy on movement of   143, 146; exclusion of   139, 140, 149, 151, 152, 155, 158; exemptions to immigration restrictions   149, 151; plantation labour   158, 161-2
commerce *see* trade
Commonwealth: health of   174; public health detention   167-8
Commonwealth Department of Health   131, 133, 134, 150-1
Commonwealth Medical Bureau   153

Commonwealth Quarantine Conference (1904)   121
Commonwealth Report on Quarantine (1912)   126
communicable disease   4, 14, 38; management of   7, 52, 83, 165, 180, *see also* contagious diseases
communication: and Empire   84; modern advances   131, 133-5; new networks in modern period   142; vaccination and vaccine matter   24-5, 36
compulsion   186, 189; and vaccination question   39-40, 41-2, 51-3, 82, 177
confinement: of Aboriginal people   98; as coerced and voluntary in sanatoria   62-3, 69, 79; of lepers   85, 86, 87, 88, 90; rationales for   60, 62, 70, 85, 86
connections   180, 185; lines of hygiene   182; made by Empire   84; made by leprosy through migration   81; monitoring of in Quarantine Station   53; of sex   4, 166, 185; vaccines   29-30
consent   7, 58, 93, 186; age of   172; to isolation in sanatoria   69; to vaccination against smallpox   50, 51, 53-4, 57
Constantinople   17; Sanitary Conference   117
constitution: Australian nation   116-17; of Commonwealth of Australia   120, 120-1; power of quarantine   116
consumptives: aestheticised idea of   64; as dangerous   64, 66-9, 77; sanatoria in early twentieth century   13, 62, 62-3, 63-5, 66, 77-80, 186; training bodies and souls of   12, 62, *see also* tuberculosis
contact   180; and epidemiology of leprosy   81, 96-7, 113-14; management of   16, 109; smallpox   15, 16, 29-30, 59; symptomlessness and the carrier   60-1; and tracking of disease   24, 29-30, 67, 130, *see also* sexual contact

contagion 4, 12, 14, 16, 180, 185; and colonialism 14, 15, 114, 152; cowpox 17, 18, 20; and cultural hybridities 107; 'dangerous' individuals 61, 152; debates about leprosy 83–93, 94–5, **97**, 98, 103, 108, 110, 113–14; and future populations 165, 183–4, 185; moral 6–7; and sexual contact 110, 113, 165; and tracking of disease 24, 67; vaccination as 19–22, 51, 128

contagious diseases: early quarantine measures 45–6; new perception of tuberculosis 65, 66, 67–8, 69; significance of smallpox 15, 35, 38, 64, *see also* communicable disease

Contagious Diseases Acts (1860s) 2, 39, 61, 69, 99, 167, 168, 169; protests 164, 167–8

contamination: and cleanliness 46–7, **47**, 48–9, 84; concerns and language of 148, 161, 162; places of 47–51; social 6–7; with vaccination from cowpox 40, 50–1

'A Contribution to the History of Leprosy in Australia' (Thompson) 96–7, **97**

Contributions to Tropical Physiology (Sundstroem) 160

convict colonies 33, 41, 119

Cook, Cecil 96, 100, 101, 105, 106, 107, 108–10, 111, 123

cordons sanitaires 2, 11, 15–16, 39, 40, 46–7, 66, 84, 87, 89, 115, 180; eugenic 13, **175**, 180–5; racial 1, 82, 103–7, 113, 114, 168

correction: institutions 70, 91

correspondence: between inmates of leper colonies 102–3, *see also Letters from a Sanatorium*

cowpox 17, 18, 20, 24, 28; and vaccination against smallpox 4, 5, 15, 16, 17–18, 19–20, 25, 36, 40, 50

Cree people 24

criminal justice: and law on vaccination 53

criminal psychiatry: diagnosis of the 'dangerous' 60, 61, 67–8, 145–6
criminality 145–6, 155, 184
Crusaders 23
Culion, Philippines 91, 101
culture: anxieties about vaccination 29; and anxieties regarding leprosy 108, 110; hybridities 107; and hygiene 5; and imagined nation of Australia 6, 136, 137, 138, 162
Cummins, S. Lyle 59
Cumpston, Dr J.H.L. 58, 68, 95, **95**, 99, 121–3, 126–8, 129, 131, 134–5, 150, 158–9, 187–9
Cyprus 86

Daily Telegraph (Sydney) 66
'the dangerous' 2, 52, 186, 187; Aboriginal people seen as 105; coloured aliens seen as 162; consumptives 61, 62, 64, 66–9, 76, 77, 79; isolation of 59–60, 70, 76, 76–7, 78, 80, 167; persistence of into twentieth-century use 170; public health detention 166–7, 172
D'Arcy Island, British Columbia 89, 92
Darwin 94, 101, 105, 131, 133
Dayman 94
Deacon, Harriet 107
Dean, Mitchell 104, 181
death: with biopolitical administration of life 181
decolonisation 2, 3
defence: military 124, 129, 135; political and biomedical borders 4, 116, 123, 137; of public health 135, 137
deportation: of Asians 4, 89, 90; leprosy laws 89, 95, 105; and segregation 158
Derby, Western Australia 111
detention: measures regarding tuberculosis 67, 69; powers regarding leprosy 54, 69, 83, 94, 98; public health 3, 11, 12, 39–40, 42, 51, 53, 83, 166–7, 167–8, 186, 187; to enforce smallpox

vaccination 50, 53, 56–7; of venereal disease suspects 166–70
diaspora: British Empire 140; Chinese 22, 81, 82, 138, 143
differences: formed by boundaries 6, 136; racial 140–1, 145
diphtheria 61, 83, 153, 178
Discipline and Punish (Foucault) 10
diseases: anxieties about vaccination 20–1, 38; carriers 60–1, 67–8, 79; circulation of 44, 113; colonial management of 12, 81–2; emergency prevention measures 11, 69; and image of Chinese men 148, 151–2; 'inherited' 174, 177, 184; marking colonial projects 24, 58; tracking of 24, 33, 135, *see also* animal disease; contagious diseases; venereal diseases
Douglas, Mary 19, 48
Dunwich, Stradbroke Island (Australia) 94, 101
Durbach, Nadja 36, 52
Dutch East Indies 131, 157

East Timor 126
Eastern Epidemiological Bureau, Singapore 134–5
economics: indentured labour 158, 161–2; and tropical medicine 157–8
edges 5, 6
Edinburgh: first British sanatorium 63
education: in Australian tropical domesticity 111; of consumptives in sanatoria 77, 78; 'dictation' test for immigrants 144–5, 149, 150; health and eugenics 7, 166, 171, 173, 177, 180, 184, 185; power exercised through 7; sex 178, *see also* training
Egypt 117, 128
Elkington, J.S.C. 101–2, 123, 159
emigrants and emigration 36–8, 146
Empire 4, 12, 16, 24, 32, 135; health of 172, 174, 188; lines of 38, 88–9; migration and movement within 140, 146–7, 151; public health detention 167–8, *see also* imperialism
Empire and Colonial Tuberculosis (Cummins) 59
The Empire Review 149, 162
enclosure 1, 5, 6, 90–1, 92, 93, 187, *see also* exile-enclosure
endemic disease 128
England 48, 66, 84, 96; anti-vaccinationism 52; cholera 119; class differentiated birth rate 172; early quarantine powers 42; hospitals for consumptives 63; philanthropy and charity regarding leprosy 83; vaccination 25–8, 33, 34, 36–8
enteric fever 61
epidemics: bureaucratic and political effect 43–5, 45; cowpox 24; information 43–5; management of 40–1, 42, 58, 119; prevention of by quarantine 125; prevention networks in modern period 133–5; smallpox 12, 22, 29, 36, 39, 41–51, 51–7, 58, 67, 122, 149; and urban spaces 12
epidemiology 9, 14, 189; Cumpston's work 123; ideas and debates about leprosy 81, 84, 95–7, 106, 108–11; and imagining of Australia 126, 130–6; information and intelligence 44–5, 134–5; observation of Nauru 'plague town' 91–2; search for origins of smallpox 23–4; significance of smallpox 15; understanding of tuberculosis 59, 65–6; vaccination 16, 34, *see also* microbes
Epidemiology of Leprosy (Cook) 96, 109, 111
epilepsy 174, 184
epileptic colonies 62, 170, 181
etiology *see* aetiology
'eugenic century' 147, 166, 172, 184
eugenics 1, 2, 4, 6, 7, 21, 79, 107, 171, 189; *cordons sanitaires* 13, **175**, 180–5; discussion of migration 145–6; distinctions made within

Index 251

white Australia 6, 104, 138–9, 140–1, 145–6, 147, 152, 153, 154, 165, 186–7; and future populations 180–5, 185; Nazi policies 165, 166; and screening 7, 13, 153–4, 155, 177; sexual conduct and regulation of population 164–6, 173–80
Eugenics Education Society 145, 180, 182, 185
Europe 125, 131; cholera epidemics 15, 117, 119, 142; health resorts 70; leprosy and isolation 69, 83; mercantilist states 7–8; open-air treatment in sanatoria 63, 64; prevention networks in modern period 133–4; search for origins of smallpox 23–4; time-distance between nations 130
Europeans: Cook's ideas of healthy community 111; and ideas about leprosy 96, 99; protection of from tuberculosis 105
exchange 5, 115, 123, 124
exclusion: of Aboriginal people 103–4; of Chinese people 88, 141; of coloured aliens 139, 140, 149, 151, 152, 155; of the 'dangerous' 60; identities and citizenship 1, 3; immigration restrictions 140, 142–4, 157; of lepers 86, 93, 102; and liberal practices 93; race-based 13, 81–2, 138, 143–7, 149; spatial 88
exile-enclosure: leper colonies 10, 12, 13, 82, 90, 91, 101–3, 186
Expansion of Races (Woodruff) 156

family history: and health 174, 177
Family Planning Association 154
Far East 130–1, *see also* Orient
'Faraway' (hulk) 56
farm colonies: for lepers 90, 91
Federal Council of Australasia Act (1885) 120
Federation Conferences 120–1
feeble-minded people 62, 166, 170, 184; arguments for compulsory sterilisation of 177, 181

feminism: concerns about venereal disease 164, 167, 171
Fenner, Frank 23
Fichte, Johann Gottlieb 5, 111
Fiji 3, 120, 125, 128
Filipinos: lepers 91
films: health education 171, 172
First World War 134, 138, 143
fitness: government concerns about population 164, 172, 173, 174, 181
folk beliefs: and fears of contagion 86, 183
foreign bodies: dealt with by quarantine service 151; and spatial governance 138, 140; vaccination as invasion by 15, 16, 18, 23, 35–6, 38; white bodies in tropics 157–63, 162–3
Foucault, Michel 8, 10, 45, 60, 82, 91, 92, 166
France: and history of passport 118–19
freedom: governance through 7, 63, 82, 104, 170, 177, 185; in leper colonies 93; of movement 69, 144
Friday Islands 94
frontiers 124, 133, *see also* interior frontiers
Future Generations 182–3

Garton, Stephen 141
gaze: medical 45, 53
genealogies: international biopolitics 137–8; liberalism 40; public health 7, 186; sanatoria 70; sexual conduct and venereal disease 164; vaccine and vaccinated children 16, 25–32
genocide 5
'geo-body' of Australia 115, 128, 131, 135–6, 137
Geographical and Historical Pathology 130
geography: asylums 87; determining of quarantine measures 123–4, 128, 130, 136; and disease 44, 158; imagining of Australia 6, 116, 125–36, 147, 161, 163; lines and

Index

segregations 6, 40, 48–9, 115, 189; racial 1, 97; and statistics 33
geopolitics: lines and segregations 6; and race 3, 9, 161
Germany: eighteenth-century medical police 42; treatment of consumptives 64, *see also* Nazi Germany
goldseeking: Chinese 59, 88–9, 147
gonorrhoea 170, 184
governance: and hygiene 1, 5, 79, 103, 165, 187; problems of detention and compulsion 40, 42, 51–7, 58; and public health 1–2, 2–3, 7–10, 10–12, 13, 33, 51, 80, 166, 186, 189; through freedom 7, 63, 82, 104, 177, 184–5
government: concerns about reproduction and population 164–5, 172, 173, 178; disciplinary 10, 91; imagining of Australia 123–4, 125, 126, 129; management of epidemics 43, 58; statistical and epidemiological knowledge 34; technologies 115, 124
Greenvale Sanatorium, Victoria 72
Gregory, J.W. 144, 157
Grey, Henry, 3rd Earl 120

Hacking, Ian 9
Hamlin, Christopher 21
Hansen, Armauer 88
Haraway, Donna 18
Harrison, Mark 30
Hawaii *see* Molokai
health: and cultivation of citizens 11, 21, 76–7, 80, 103, 116–17, 166, 173–4, 178–80, 189; documents 119; of Empire 83–4, 172, 174, 188; and eugenics 155, 177–8, 185; and identity 1, 4, 13, 49, 162; interwar ideas and policies 3, 5, 111, 137–41, 146, 156–7, 163, 164, 166, 173–4, 180, 189; nineteenth-century conceptions 30; of populations 7–10, 13, 33, 42, 58, 115, 116–17, 147, 178, 184, 189
Health (journal) 133
Health and Empire 170–1, 173
Hennessy, John Pope 143
heredity: and infection 21, 184; and leprosy 83
Hislop, Gordon 75
historians 1, 2, 165
History of Sexuality (Foucault) 166
HIV/AIDS 77, 124–5, 170, 187
homeopaths: interest in vaccination 19
homosexuality 145–6, 184
Hong Kong 83, 143, 151
hospitals: in British colonies 86; for consumptives 63; infectious disease 47–8, 94, 101, 155, 170; isolation 61, 66; women's 173, *see also* Lock Hospitals
Hudson Bay Company 24
Hunter, Ernest 98
Huntington, Ellsworth 159
Hutt, C.W. 142, 143
hybridisation: programme of 109
hybridities: cultural 107; of sanatoria 13, 167
hygiene 4–5; and citizenship 62, 77–9, 173; and governance 1, 5, 79, 103, 165; and immigration 142–3, 151–7, 160–1; imperial 5, 12–13, 113, 165; international 5, 13, 134, 137–8, 140, 141–7, 148, 161, 162, 163, 165; lines of 1, 2, 10–12, 13, 16, 36, 38, 40, 48–9; racial and national 2, 5, 6, 7, 103, 123, 138–41, 144, 147, 151–7, 158–63, 164, 165, 189; sex 165, 166, 171, 178, 185; social 164, 166, 171, 172, 173; tropical 5, 111–12, 157–8, 159; vaccination and crossing of boundaries 20, *see also* mental health and hygiene

Iceland 81
identity: formed by boundaries 6, 136; and health 1, 4, 13, 49, 135–6, 154, 180; national and racial 116, 147, 148; significance of security 135; white self in tropical Australia 111
Illustrated Sydney News 47

imbecility 145–6, 153
immigration: concerns about incoming ships 131; and Cook's theory of leprosy 108; eugenic ideas 156–7, 182, 185; and international hygiene 142–3, 151–7, 160–1; laws and regulation of 88, 89, 95, 105, 111, 123, 124, 129, 134, 135, 146, 148, 149–50; as racialised 5, 6, 13, 116, 137–41, 142–4, 147, 150, 155, 162, 182; restriction lines 1, 7, 13, 103, 115, 136; shaping of populations 145, 145–7, 162
Immigration Restriction Act (1901, later Immigration Act) 95, 105, 111, 137, 143–4, 147, 150, 150–1, 152, 153, 154, 159
immune system 18
immunisation: mass 2
immunity 4, 18–19, 35, 36, 74, 96
Imperial Social Hygiene Congresses 171
Imperial Vaccination Acts 2, 13, 34, 51
imperialism: anxieties about racial exclusion 143–4, 162; and leprosy 81–2, 83–4, 87, 88–9, 93; making boundaries 6, 89; and problem of venereal disease 171; tropical medicine 140, 157–8, 188; and vaccination 15, *see also* Empire
imprisonment 52, 98, 169
indentured labour 59, 88–9, 108, 137, 140, 148, 151, 158
India 3, 28–9, 30–2, 40, 117, 125, 128, 131, 168; leprosy 83, 88; practices of inoculation 16, 20
Indian people 143, 149
Indigenous people: Australian colonisation of 4; effect of diseases on 14, 15, 41; 'management' of 141, 149; public health and colonial administration 9, 81, 149; relations with British-whites in Australia 6, 82; seen as 'dangerous' 60, *see also* Aboriginal people

industrial settlements: for lepers 90, 91
industrialisation 8, 64, 79
infantile paralysis 61
infection: and the 'carrier' 60–1; circulation of through vaccination 16; epidemiologial intelligence 134–5; and heredity 21, 184; immunity achieved through 18, 19; racialisation of leprosy 99, 104; and racialised isolation of Aboriginal people 105; and resistance 74; and sex 178
infectious disease: Australia seen as free of 126; Australia's vulnerability to 116, 131; and ideas about Aboriginal people 105; maps 125, 126, **127**; places of segregation 39, 40, 47–8; preventive quarantine measures 117–21, 182; prohibition of sufferers 153; and trading relations 42
Infectious Disease Supervision Act New South Wales (1881) 39, 42
Infectious Diseases (Notification) Act (1889) 61
influenza 121, 153, 167
information: collecting 8–9, 62; epidemic as 43–5; modern technological advances 131, 133–5; and vaccination 33–4
inoculation 15, 16, 16–17, 119, *see also* 'auto-inoculation'
inspection 124; of immigrants 139, 142, 152–3; of imperial and global movement 137, 142; of venereal disease suspects 167, 170; of vessels and of people 141–2, 150, 153
institutionalisation: colonial systems 82, 186; of epidemiology 9; lines and segregations 6; in management of leprosy 81, 87, 93; and the nation-state 116, 118–19, 121–3; open-air treatment in sanatoria 64, 66, 70, 75–6, 77, 79, 80
interior frontiers 5–6, 6, 107–13

International Biological Programme
(1960s)   14
International Hygiene (Hutt)   142, 143
International Leprosy Congress
(1897)   88
international relations   135, 147
internationalism: development of
   134, 142
Invalid Pension Act   78
invasion   4, 116, 129; anxieties
   about sex and race   108, 113; and
   borders   123, 137; by disease   59,
   129; vaccination as   15
The Invention of the Passport (Torpey)
   118–19
Irving, Helen   125
Isaacs, Isaac   148
island-nation: Australia as   13, 116,
   124–8, 129, 130, 136
islands: leper colonies   80, 82, 94,
   98, 101–3, 186; lock hospitals   169
isochronic charts   131, **132**
isolation   7, 189; of Aboriginal people
   168–9; anti-vaccinationist activity in
   Leicester   22; coerced   69, 86;
   compulsory   51, 80, 90, 104, 106;
   of consumptives   62, 66–9, 76, 77,
   79, 105; of the 'dangerous'   59–60,
   70; as emergency response to disease
   11; eugenics   181, 185; legislative
   and management powers   61, 62,
   80; management of leprosy   81, 82,
   83, 88, 90, 91, 94, 98, 100–2, 103,
   106, 182–3; natural   14; in
   quarantine   129–30; racial policies
   22, 39, 40, 41, 56–7, 67, 94, 98,
   99–100, 103; smallpox quarantine in
   New South Wales   39, 50, 54, 66;
   therapeutic   79–80, 86; voluntary
   58, 59, 62, 62–3, 69, 76–7, 79, 80,
   87, 90, 187

Jamaica   83, 85, 85–6
Japan   111–12, 122, 143, 151
Japanese people   6, 143, 156, 157,
   161; pearlers   149
Java   89
Jenner, Edward   17, 25, 28–9, 30
Jews   1817

Kanakas *see* South Sea Islanders
Kern, Stephen   133
Kimberley, Australia   98, 105
knowledge: development of
   techniques   8–9, 38
Koch, Robert   19, 64, 65, 69, 88
Koepang   151

Labor Party   147, 148, 177
labour *see* indentured labour
Lambert, Agnes   84, 86
The Lancet   50
laws: and compulsory vaccination
   53; control of leprosy   81, 88, 93,
   93–4, 99; immigration   138, 140,
   143, *see also* legislation
lazarets/lazarettos   87, 88, 94, 100,
   104, 106
League of Nations   122, 134, 142, 143
legislation: anti-Chinese   149;
   quarantine   119–21; to contain
   infectious diseases   39, 61, 98,
   117–23; vaccination   33, 51, 53;
   venereal disease   168, 169, *see also*
   laws
Leicester: anti-vaccinationist activity
   22
leper colonies   62, 80, 84, 87, 88,
   91–3, 170; exile-enclosure   10, 12,
   13, 82, 90, 91, 101–3, 186
'Leper Line': Australia   6, 97, 106–7,
   161
lepers: British colonial institutions for
   90–1; contagion and segregation
   83–93; as 'dangerous'   80; Foucault
   on treatment of   10, 82, 91; island
   isolation in Australia   93–103;
   isolation of   69, 86–8, 91–3, 182–3
leprosaria   88, 98
leprosy   21, 41, 74, 153, 181; in
   Aboriginal people   12, 94, 95, 96,
   98, 99, 100, 103, 104, 105, 106,
   108–9, 113; as an 'imperial disease'
   81–2, 83–4, 87, 88–9, 93; Australian
   isolation laws and policies
   93–103, 106; and Chinese people
   88–9, 95, 96, 99, 108, 110, 142–3,
   148; confinement   39, 86, 87, 88;
   contagion and segregation   83–93,

106; debates about contagion 83–93, 94–5, **97**, 108, 110, 113–14; debates about segregation and isolation 89–91, 99, 170; detention powers 54, 67, 69, 94, 98; notification of 61; policy and management of 6, 12, 13, 58, 59, 81–2, 88, 98–103, 107; racial distribution 83, 95–7, 103–7; sexuality, contact and race 107–13
Leprosy Acts (1890s) 69, 88, 94
Leprosy Commission (1891) 88
Letters from a Sanatorium 75–6, 78
Levine, Philippa 168
liberalism: arguments against isolation-as-prevention 69; debates on compulsion 52, 93–4, 104, 170; genealogy 40; governance and public health 3, 7–10, 13, 39–40, 61, 62, 115, 122, 170, 177, 189; and self-regulation 76–7, 93; shifting modes of rule 2–3
liminal places/spaces 40, 129
lines or barriers 2; quarantine 136, 137, 182; time-distance between nations 130, 133, *see also* borders
Lister Institute of Preventive Medicine 122
Lock Hospitals 6, 61, 167, 168, 186
London 22, 25, 32, 130, 143, 152
London School of Hygiene and Tropical Medicine 96
lunacy 42, 166, 180
Lunacy Acts 167
Lupton, Deborah 11, 103

Macassar 151
Macau 83
McClintock, Anne 110
McGregor, Russell 141
Mackellar, Dr Charles 118, 120
Maclean, L.H.J. 23
Madagascar 83
Madras 83
Malay Peninsula 128
Malay States *see* Sungei Buloh
Malays: forced onto leper colonies 94

Manderson, Lenore 135
Manson, Patrick 81
maps and mapping 125, 126, **127**, 130, 131, **132**, 136
maritime quarantine 7, 13, 15, 22, 45, 116, 117–18, 119–20, 124, 125–6, 136, 137, 150–1
marriage: idea of health screening for 174–7, **175–6**, 177–8, 184; promotion of 172
Marx, John 21
Mauritius 83
Mayne, Alan 46
Mecca: annual pilgrimage from India to 117
medical history: Cumpston's work 123; on prevention 60; smallpox 15, 23–4, 59; vaccination 16
Medical Journal of Australia 94
medical police 42, 172
medicine 2, 32, 34; and penal systems 53, 67–8; state or social 7, 40–1, *see also* tropical medicine
'medieval'/premodern systems 82, 86, 90, 98–9
Melbourne 20, 126
men: sexual conduct of 178
mental deficiency 166, 174; and asylums 181; and sterilisation 181–2
mental health and hygiene 172; eugenic ideas 155, 184; immigration law and regulation 146, 152–3; sanatorium regime 75
mercantilist states 7–8, 119
metaphors: of invasion 15; the protective net 124, 125, 126; of public health and hygiene 7, 183, 184; 'seed and soil' 14, 74, 79, 178, **179**; the 'social body' 4, 34
Metchnikoff, Ilya 18–19
Mexico: US border with 124
microbes: and air travel 133; asserting origins of 24; early ignorance of 45; studies 158, *see also* epidemiology
middle classes: consumptives 64, 66, 69, 70, 72, 75–6
migrants 2, 14

migration: and connections made by leprosy 81, 88–9; imperial and biopolitical 88–9, 146–7, 151–7; and interwar ideas about infectious disease 105, 137–41, 145–6, 156–7, 182, 184; regulation of 119

the military: in British colonies 14; Commonwealth Army 129; first arrivals in Sydney (1788) 41

military-colonial discourses 14, 129, 135

miscegenation 107, 110

missions and missionaries 14, 25, 87, 105

modern period: infectious disease segregation 39, 79; leprosy management 82; national and colonial networks 142; public health management 186; technology of quarantine 131–3

Molesworth, E.H. 96, 99, 106

Molokai, Hawaii 87, 92

monitoring *see* surveillance

Montague, Lady Mary Wortley 17

Montreal 128

moral concerns: health and cleanliness 21–2; image of Chinese men 148; sexual conduct 178; vaccination 22

morbidity 43, 44

Morin, Edgar 124

mortality: infant and maternal 173; tuberculosis 59, 65–6

movement: association of vaccination with 15, 16, 38; between nations and continents 117, 153; biopolitical restriction of 140; imperial and global 137, 146–7, 156; monitoring of in Quarantine Station 53; regulation of by documents 118–19; regulation of in management of leprosy 81; restriction of Aboriginal people 104, 105, 113

Muir, Dr Ernest 102

Muslims 117

Natal 144

natalism 172

nation and nationalism 3, 4, 9; boundaries of rule 13; Commonwealth of Australia 116, 119–23, 129; eugenics and race 141; as 'geo-body' 115, 136, 137; health and identity 4, 111, 116–17, 123, 135, 147, 162, 178, **179**; ideas during interwar period 3, 137–41, 146, 180, 189; and institutionalisation of public health 116, 118–19, 121–3; and medicine 2; and quarantine 116, 117–23, 129, 135; race and hygiene 2, 3, 5, 6, 7, 103, 123, 136, 138–41, 144, 148, 151–7, 158–63, 180, 187–8, 189

National Emergency Act (1942) 170

National Health and Medical Research Council (NH&MRC) 100, 104

National Socialism 2, 181

National Vaccine Establishment, London 25

Native Administration Act (1941) 106–7

Nauru, Pacific 91–2

Nazi Germany 2, 138, 177, 178

Nazis: enclosure of Jews in Warsaw 7; eugenic policies 165, 166, 178

neoliberalism 93

neurasthenia 70, 111

New Brunswick 81, 83, 87

New Guinea 157

New Poor Law 33, 52

New South Wales 18, 120, 149, 154; Board of Health 39, 42–3, 54, 57–8; bureaucratisation of health 39, 42, 43–5; Chinese goldseekers 147; Committees on Vaccination 21, 28, 32; Eugenics Society 174; immigration exclusion measures 143, 144; Infectious Diseases Hospital 94; information on leprosy 83, 96–7; interwar campaign for health and responsibility 178–80, **179**; Leprosy Act (1890) 69, 88, 94; Motherhood Endowment Scheme 173; notification of infectious

diseases 61; Prisoners' Detention Act (1908) 169; Royal Society 118; smallpox and vaccination 23, 24, 25, 33, 51–2
New York: 'Typhoid Mary' 61
New Zealand 83, 120, 125, 143, 144
The Nineteenth Century 84
Norris, W. Perrin 127
North Africa: French imperialism 157
North America 15, 41, 63, 138, 155, *see also* Canada; United States of America
Northern Territory 147
Norway: leprosy 83, 88
notification 61–2, 83; Australian leprosy laws 94; Australian tuberculosis measures 67; policies in Leicester 22; venereal diseases 169

Office international d'hygiene publique 133–4
open-air treatment *see* sanatoria
Orient 117, *see also* Far East
'others': exclusion of 5
Ott, Katherine 64
Ottoman Empire 16

Pacific colonies 25, 120, *see also* Nauru
Pacific Island Labourers Act (1901) 137
Pacific Islanders 149, 157, 158
Paddington, Marion 174
Pan-American Sanitary Bureau 142
pandemics: AIDS 124–5; influenza 121, 153
Panopticon 92
Papua New Guinea 126
Parkes, Sir Henry 148, 149
passport 118–19, 142; vaccination 36
Pasteur, Louis 19
Pasteur Institute 122
pathology: of attributes 184; and race 49; of sex and reproduction 166, 172
Peel Island 94, 99, 101–3

penal systems: and detention of lepers 98; dovetailing with medical systems 53, 166; and segregation 39, 68, 181, *see also* punishment
penicillin 170
peripheries 5
Perth, Australia 1126
Petersen, Alan 93, 103
philanthropic concerns 7, 22, 33, 83, 86
Philippines 122, 128, 157, *see also* Culion
plague 94–5, 98, 117, 118, 126, 131, 135, 153; in Australia 58, 129; 'great white' 59; in India (1896) 40; isolation measures 65, 67, 167; notification of 61
'plague towns' 10, 15, 91–2
Police Offences Act 169
policing 186; and detention of lepers 92, 98, 104; during cholera epidemics 15; implementation of *cordons sanitaires* 39, 105–6; public health police 42–3; quarantine as 123; and self-policing in sanatoria 70, 76, 80
polio 61
political economy 7, 9, 22, 115
politics: compulsory vaccination 52; debate on sexual conduct and venereal disease 164; of epidemic 43; hygiene 5; in origin-sourcing 24; public health 7–10, 165; racial aspirations of Australia 137, 187; self-monitoring in sanatoria 77; white race and the tropics 159, 161
the poor 8, 9
Poor Law *see* New Poor Law
Poovey, Mary 9
populations: bureaucratic effect of epidemic 43–5; concept of 9; contact and infection between 15, 21; effect of tuberculosis 59; eugenics 164–6, 180–5; health of 7–10, 13, 33, 42, 58, 115, 116–17, 147, 178, 184, 189; and immigration 145, 145–7, 156,

162; New South Wales 51–2; nineteenth-century racial conception of 30; pure and naturally isolated 14, 16; reproduction of 172–4, 180; as social body 4, 8, 9, 33, 34, 42, 44, 166

Porter, Dorothy 8

ports: and determining quarantine measures 130–1, *see also* maritime quarantine

postcolonialism 3, 4

power: and disciplinary government 10, 91; and Foucault's view of punishment 8, 10; public health and liberal governance 7–8

prevention: confinement and isolation 60, 69, 70; and contamination 51; debates about leprosy 84, 87, 88, 89; and eugenics 180, 181; legislation 39; new networks in modern period 133–5; and smallpox vaccination 22, 40; tuberculosis 62, 63, 64–5, 69, 77, 79

preventive geographies 1

prisons 60, 155, 169

Proctor, Robert 7

propaganda: health 166, 171, 173–4, 178–80, 184

prostitutes 2, 5, 99, 108, 170, 184; Lock Hospitals for 61, 167, 168

psychological discourse: sanatoria 75, 80

public health: boundaries of rule 13, 110; British 3, 10; bureaucratisation of 2, 12, 36, 39, 40–1, 42, 43–5; centrality of quarantine 116, 121, 162; and colonial governance 9, 9–10, 13, 14, 188–9; detention 3, 11, 12, 39–40, 42, 51, 53, 56–7, 166–7, 187; and eugenics 165, 166, 180, 184, 185; genealogies 7, 186; and governance 1–2, 2–3, 7–10, 10–12, 13, 33, 51, 166, 186, 189; and imperial policy 171; and national identity 4, 111, 116–17, 123, 129, 135–6, 178; nineteenth-century 'improvement' arguments 21–2; problem of leprosy 81, 95; and race 1, 3–4, 6, 9, 95, 105–7, 109, 110, 114, 136, 137, 142, 163; and segregation 2, 3, 11, 12, 22, 39; significance of tuberculosis 65, 66, 70; spatialised governance 9–10, 33, 42–3, 59–60, 62, 80, 109, 113, 180–1; 'the new public health' 10, 10–11, 79, 186

Public Health Act (1888), Australia 94

puerperal fever 61

punishment: Foucault's view 8; implementation of *cordons sanitaires* 39; for refusing smallpox vaccination 50, 52–3; and treatment of lepers 101–2, *see also* penal systems

purity: biological and racial 14, 32, 105–6, 109, 162; connected with sex and reproduction 165, 166; imagined community of Australia 5, 13, 110, 111, 128, 131, 137, 138, 154, 162; and resistance to disease 74, 128; and vaccination 12, 29, 35–6, 37, 128

quarantine: abandonment of for new public health 11, 79, 128, 186; boundaries of rule 1, 6, 12, 40, 103; as central to Australia 116, 121, 124–5, 126, 128–9, 135–6, 150–1, 162, 188; early systems 10, 66, 129; emergency measures for cholera 167; isolation in 129–30; leprosy laws 89; mapping and space-time delineation 130–6; and national borders 6, 115, 123–9, 135–6, 142; and nationalism 116, 117–23, 129, 135; and new Australian Constitution 116, 120–1, 125; powers 116, 121–3, 142, 153; and racialised immigration 137–41, 148, 151, 182; and sanatoria 66, 79; smallpox 36, 40, 41, 42, 45–51, 53–7, 65, 149; stations 39, 80, 84, 88, 94, 98, 125–6, 129

Quarantine Act (1832)   119
Quarantine Act (1908)   94, 111, 121, 133, 153
Quarantine Station, Sydney   39, 40, 41, 47, 48–51, 53, 54–7, 101, 129
Queensland   120, 125, 131, 149, 157, 170; anti-Chinese agitation 147; immigration exclusion act 143; indentured labour   158; Leprosy Act (1892)   69, 88, 94; leprosy and treatment of lepers 95–6, **95**, 97, 99, 100–2, 105, 108; lock hospital system   168
Queensland Health Act (1911)   168

race: and Australian management of leprosy   82, 83, 95–7, 99–100, 103, 114; biologising of   30, 32, 145; Cook's theory of transmission of leprosy   108; and creation of Australian nation   136, 137–41; delineation of   13; discrimination against Chinese   94; and eugenics 140–1, 153–4, 164; and exclusion 13, 81–2, 138, 143–4, 149; as factor in 'dangerousness'   60; ideas about im/purities   14, 32, 105; and ideas about tuberculosis in USA   64; ideas during interwar period   3, 109, 111–12, 137–41, 145–6, 154, 158, 163, 180, 189; and imaginings of white Australia   147–51, 162; and immigration restrictions   6, 13, 116, 137–41, 142–4, 147, 151–7, 162, 182; and making boundaries 6, 110; and national hygiene   2, 5, 6, 7, 103, 123, 138–41, 144, 147, 151–7, 158–63, 189; and nationalism   2, 3, 6, 136, 148, 162; and public health   1, 3–4, 6, 9, 95, 105–7, 109, 110, 114, 136, 137, 142; and sexual contact   30, 82, 105–6, 107–13, 166; susceptibility theories   91, 92; theories of immunity   96; and vaccination 17, 29; welfare policies   173, 189; and whiteness of Australia   139, 141, 150, *see also* racial segregation

Race Improvement Society   171
Racial Hygiene Association (New South Wales)   154–5, 156–7, 171–2, **175–6**, 180
racial segregation   2, 3, 6, 11, 13, 49; *cordons sanitaires*   1, 82, 103–7, 113, 114; isolation strategies in Australia   57–8, 99–100, 105–7; spatial strategies   104, 109; through immigration acts   158
Red Cross Societies   134
reform: of consumptives in sanatoria 70, 72–4, 80, 186
religious concerns: Christian discourse 82; leprosy 83; vaccination   22
Report on Consumption (1911), Australia)   66–7, 68
Report on Quarantine… (Norris)   **127**
A Report to the President of the Board of Health… (Thompson)   37
reproduction: and eugenic ideas   164, 173–80, 184; and responsibility 164–5, 172, 178–80, **179**; and sexual conduct   8, 165, 180–1, 185
resistance   4, 116, 128; through open-air treatment   74–5
Richards: Thomas see *Animal Vaccination*
risk: strategies of prevention   60, 186
Robben Island   81, 87, 92, 107
Rockefeller Foundation   122, 134, 142
Rogers, Sir Leonard   89–90, 93–4, 100, 170
Roman Catholic Church   177
Rose, Nikolas   7, 76, 77, 189
Rosen, George   7–8
Rosenberg, Charles   40
Royal College of Physicians   83, 84, 85, 88, 99, 107
Royal Commission on Small-Pox (1883)   46, 54–7
Royal Society, New South Wales   118
Royal Vaccine Institute   29

St Lucia   85
San Francisco   22
sanatoria   13, 59, 62, 62–3, 63–5, 66, 170, 186; for Aboriginal people 104–5; as carceral spaces   68, 167;

coerced and voluntary confinement 59, 62–3, 69, 79; and cultivating healthy citizens 12, 13, 62, 70–9, 80; as modern quarantine 66, 79
Sanatory Camp, Little Bay (New South Wales) 41, 45, 47–8, 48, 49, 53, 54, 94
Sanitary Act (1866) 22
Sanitary Conferences: Australasian (1884) 108, 126, 128, 130, 135; international 118, 124–5, 133, 138, 142
sanitary reforms 22, 33; quarantine measures 117–23; regulations 126
sanitary science 10, 34
sanitation 1, 134, 135, 188
SARS 187
Saunders, Suzanne 104
Scandinavia: compulsory sterilisation 165, 177
science: racial ideas during interwar period 3, 109, 111–12, 137–41, 145–6, 154, 158, 165, *see also* sanitary science; social sciences
seamen: and circulation of diseases 2, 142
Sears, Alan 11
Second World War 2, 4, 134, 137, 142
segregation 7, 186, 187; in administration of epidemics 40; borders 38; consumptives in sanatoria 64, 66, 67, 68, 69, 79, 104–5; and contagion 14; eugenics 139, 166–7, 180–1, 182, 184, 185, 186; leprosy and leper colonies 13, 82, 83–93, 99, 100–3, 103–7; Lock Hospitals 6; places of infectious disease 39; in public health 2, 11, 12, 22, 39–40, 189; rationales for 84, 85, 167; rigidity of in Australia 107, 122; spatial strategies 46, 79, 184; venereal disease management 166–7, 168–9, *see also* racial segregation
self: cultivation of in sanatoria 62, 70–7; cultivation of whiteness 111, 113; health and identity of 4, 11; and the immune system 18; interventions in governing of 164, 186; vaccination as invasion of 15, 18, 23, 38
self-governance 13, 59, 62, 79, 99, 104, 166, 186, 188
self-surveillance 11, 53, 76–7
settlement: association of vaccination with 15; Australian 3–4, 111–12, 159, 161; nineteenth-century idea of colonies 155; and racial exclusions 138
sex and sexual conduct: biological connections 4, 141; connection with politics and power 8; education and hygiene 171–2, 185; and image of Chinese men 148; and regulation of population 164–6, 172, 178, 180–1, 185; and venereal disease 167
sexology 164, 172
sexual contact: and making boundaries 6; policing of 105–6; and race 30, 82, 105–6, 107–13, 166
ships: concerns regarding quarantine 131, 134; control of 115, 141–2, 153, *see also* maritime quarantine
Siam Mapped (Thongchai) 115
Sibley, David 5
Singapore 131, 134, 151
smallpox 2, 15, 64, 94–5, 98, 113, 126, 131, 135, 148, 153, 186; documentation of procedures 33, 38; global and colonial tracking of 24; as 'invading' Australasian colonies 59, 129; Jenner's views 28; quarantine measures 6, 12, 36, 39, 48–9, 128, 149; scars 35–8, **37**; search for and ideas of origins 23–4, 59; in Sydney 12, 29, 36, 39, 40, 41–51, 51–7, 58, 69, 122, 149, 167; vaccination against 15, 16, 16–23, 35–8, 40, 46–7, 50–1, 128–9; vaccine genealogies 25–32
Smart, Judith 170
Smith, Adam 9
Smith, Kingsford 133
Smith, Ross 133

social policy: development of 8, 165
social sciences 8, 22
social spaces: leprosy management 82; re-ordering of 45, 48–9
sociology/sociologists 1, 9, 10, 11, 59–60, 79, 165
soldiers 2, 169, 170
souls: training of consumptives in sanatoria 12, 62, 73, 74, 77, 100; Victorian idea of cleanliness 5
South Africa 110, 144, 182, *see also* Robben Island
South America 138
South Australia 120, 143, 147
South Pacific Islanders ('Kanakas') 12, 94, 99, 108, 110, 148
Soviet Union: propaganda style 178
spaces: heterotopic 62; imagining Australia 129–36, **132**; imperial 84; of isolation 83; public health 1–2, 12, 43, 53, 62, 186; quarantined 45, 50, 79, 89, 104, 123; racial 105, *see also* carceral spaces; therapeutic spaces; urban spaces
spatial management: of consumptives in sanatoria 71–2, 79, 80; of Indigenous peoples 81; of public health 43, 82, 105–7, 109, 113, 138, 180–1; racial *cordons sanitaires* 103, 105–6; Victorian modes of 33
species: and vaccination 20, 23, 29
Springthorpe, J.W. 67–8
statistics 8–9, 22, 33, 34, 44–5, 49, 155, 172–3
Stein, Gertrude 133
sterilisation: compulsory 165, 172, 177, 181; and eugenics 181–2, 184, 185
Stern, Alexandra Minna 124
Stoler, Ann Laura 5–6, 107, 113, 166
Stone, Dan 141
Stopes, Marie 154
Story of Medicine (1954) 23
Straits Settlements 3, 83, 131, 157
subjectification 6, 7; nationality and race 147; training of lepers 91, 93

Sudan 124
Suez Canal 117
sugar industry: Queensland 158
sulphonamides 170
Sundstroem, E.H. 160
Sungei Buloh: leper settlement 92–3, 101
surveillance: of borders 123–4; epidemiological 135; management of venereal disease 167; medical 42, 53; national systems 142; of Nauru 'plague town' 91–2; Panopticon 92; in sanatoria 71–2, 78
swine-pox 28
Sydney: Australasian Sanitary Conference (1884) 108, 126, 128, 130; Health Week conference 181; in imagining of Australia 126; infectious diseases hospital 101; lines of hygiene 40; municipal policy on tuberculosis 65; public health isolation 22; smallpox epidemics 12, 29, 36, 39, 40, 41–51, 51–7, 58, 122, 149
Sydney Morning Herald 46
syphilis 20, 32, 83, 95, 107, 170, 183, 184
Syria 86

Tasmania 63, 83, 120, 143, 144
Tauber, Alfred 19
Taylor, T. Griffith 159
technologies: communications 131, 133–5; of government 115, 124; quarantine 131–3, 162; vaccination 24, **26**, 38
Teller, Michael 63
theosophist radio station 2GB 172
therapeutic spaces 1, 45, 86, 91
therapies: for neurasthenia 70; new public health strategies 62; treatment of tuberculosis 63, 65, 70, 72–4, 79–80
Thompson, J. Ashburton 37, 96–7, **97**, 98, 128
Thongchai, Winichakul 115
Those Who Come After 182, 183
Thursday Island 125, 126, 131, 151

time: imagining Australia in 129–36, **132**
Torpey, John 36, 118–19
Townsville 125, 159, 160
trade 2, 24, 42, 118, 119, 123, *see also* maritime quarantine
trade unions 147
traffic: and hygiene 142
training: in leper colonies 91; in sanatoria 62, 70, 72–3, 78, 79, 80, *see also* education
travel: modern advances 131; restrictions on Aboriginal people 107; and vaccination 16, 36, 38
Trinidad 83, 84
tropical diseases 74, 81, 84
tropical medicine 14, 83, 105, 122, 140, 156, 157–63
tuberculosis 59, 88, 95, 153, 155, 177, 183, 187; changing ideas of 64, 66, 79, 83; isolation of 'dangerous' consumptives 62, 64, 66–9, 98; management of 58, 59, 61–6, 67, 69, 79, 80, 104–5, 113; sanatoria 12, 41, 63, 70–7, 104–5
typhoid 61, 153

United States of America 144, 177; border with Mexico 124; change in ideas about tuberculosis 64; colour bar 182; immigration exclusion laws 143, 146; sanatorium treatment 63; yellow fever 117, *see also* San Francisco; Washington DC
upper classes: open-air treatment for consumptives 64
urban spaces 2, 12, 43, 115; sanatoria for consumptives in early twentieth century 13; segregation of African-Americans in USA 64
urbanisation: and tuberculosis 64, 79
Utilitarianism 8
utopias 6, 180

vaccination: administration of 33–8, 189; of animals at Bombay 26; arm-to-arm method 34, 35; and colonialism 15, 16, 25–6, 38; compulsory 3, 7, 12, 17, 39, 41–2, 50–1, 51–3, 98, 128–9, 167; cowpox 4, 12, 16, 17–18, 19–20, 36, 40–1, 50–1; documents 119; genealogies 16, 29–32; smallpox 15, 16, 16–23, 35–8, 40, 41, 46–7, 50–1, 128–9; Victorian debates 19–23
Vaccination and its Evil Consequences (Morton) 32
vaccines: communication of matter for 24–5, 29; genealogy 25–9
variola discreta: photograph of scars 37
variolation 15, 16
Vaughan, Megan 91
venereal diseases 2, 61, 109, 153, 155, 173, 174, 184; detention and education 98, 154, 166–72; feminist debate 164; Lock Hospitals 6, 61, 167, 186
Victoria, Queen 120
Victoria, Australia 36, 61, 83, 94, 97, 120, 147; Eugenics Society 177–8, 183, *see also* Greenvale Sanatorium
Victorian period: colonial contexts 9; culture of cleanliness 5; global lines of Empire 38; liberalism 2–3; social science 8, 22; vaccination debates and anxieties 15, 19–23, 29
Virchow, Rudolf 19

Wales 33, 63
Walker, David 161
Warsaw: Nazi enclosure of Jews 7
Washington DC: International Sanitary Conference (1881) 118
Watson, J.C. 148
Weihen, Dr A. Wallace 139
welfare 3, 8, 64, 117, 122, 165, 172, 173, 189
West Australia 120, 168–9
West Indian colonies: leprosy 83, 85
West Indies 144, 157
Western Australia 125, 143, 144
The White Man in the Tropics (Cilento) 105, 159

white people: and Australian national policy 3, 4, 123, 126, 128, 137–41, 140, 145, 149, 156, 159; classification of for quarantine measures 6, 49, 50; and eugenics 6, 104, 138–9, 140–1, 145–6, 147, 152, 153, 154, 182, 186–7; and governance-through-freedom 63; leprosy in 81, 83, 89, 95–6, 97, 98, 102–3, 112; and racial imaginings 89, 110–13, 147–51, 154, 158–63; settler societies 1, 3–4, 110–13, 138–9, 186–7; and smallpox in Australasian colonies 59; and tuberculosis in Australia 58, 64, 69, 77, 79, 105; with 'undesirable' characteristics 13, 60, 152, 153
whiteness: management of 2
Wilkinson, H.L. 182
Wise Parenthood (Stopes) 154

women: compulsory detention and examination of 167, 168; and concerns about reproduction 172–3; and Cook's theory of leprosy 108; encouragement of whites to tropics 110–13; and eugenic ideas 164; traffic in 142
Woodruff, Charles Edward 156
Worboys, Michael 87, 89, 90
workhouse: tradition of 66, 91
working classes: anti-Chinese feelings 147–8; anti-vaccinationism in England 52; management of 33; reform institutions 70; and tuberculosis 64, 66, 67, 69; understanding of vaccine scar 36
World Health Organization 134

yellow fever 117, 118, 131, 153
Young, W.J. 159, 160

Lightning Source UK Ltd.
Milton Keynes UK
UKHW022139091019
351276UK00016B/420/P